# FUTURE CLASSICS

Shamal

# FUTURE CLASSICS

**PRESENT-DAY CARS DESTINED TO BECOME TOMORROW'S CLASSICS**

ALEXANDER CORNE

B. Mitchell

A QUINTET BOOK

First published in Great Britain in 1991


Produced for W.H. Smith Limited
by Quintet Publishing Limited

ISBN 0-88665-948-5



This book was designed and produced by
Quintet Publishing Limited
6, Blundell Street
London N7 9BH

Creative Director: Terry Jeavons
Designer: Wayne Blades
Project Editor: Sarah Buckley
Editor: Rosemary Booton
Picture Researcher: Mirco Decet

Typeset in Great Britain by
Central Southern Typesetters, Eastbourne
Manufactured in Singapore by
Chroma Graphics (Overseas) Pte. Ltd.
Printed in Hong Kong by
Leefung-Asco Printers Limited

# CONTENTS



NOTES TO THE TECHNICAL SPECIFICATION BOXES: *Where the launch price is separated by a semi colon, to the left of the semi colon is the UK launch price and to the right the US launch price, except in the case of US-built cars, where the US launch price is quoted first.*

# INTRODUCTION

Picking the classics of tomorrow from the cars of today was no easy task. For, although present-day commentators bemoan the lack of cars bursting with character in the era of the nondescript tin box, there is however a saviour for the motoring enthusiast in the 'niche' car.

Today, as much as in any time past, people with large sums of money can afford to be different. So the motor manufacturers have devised ways of attracting attention with the evolution of 'concept' vehicles, designed to stir the imagination. Putting concepts into limited production brings the cars to life, and retains that vital spark of interest in the marque itself, or maybe simply in one particular model range. Rarity in these cases is a major factor.

LEFT

*Only six Bugatti Royals were ever built; Ettore Bugatti intended them for the royal families of Europe only. Surprisingly, however, none of the intended customers were interested and three of the six cars were given to members of the Bugatti family.*

Ten years ago, classic cars were not seen merely as old motoring relics, but rather as the last remnants of a bygone 'golden' age of motoring. They had that hard-to-define, but almost tangible, aura of character and desirability, something modern volume manufacturers mistakenly think of simply as 'wood and leather' interiors.

For the established classic car enthusiast, a car built in numbers less than 100 makes for an interesting curiosity, a car worth preserving due to its rarity. Between 100 and 1,000 examples built and the car is still worth pursuing, but often a less wise investment. Cars built in their thousands cannot ever hope to compete with the Bugatti Royales of this world, or even the next.

The 'popularization' of classic cars during the late 1980s led to a mass scramble to acquire a classic of any sort. Hence the rush to buy up Austins, Morris's and Fords; cars which in their day did a reasonable job of work, but which generally lacked any special qualities which could label them as classics.

LEFT
*The Lotus Seven debuted in 1957, a two-seater roofless, doorless, roadgoing fun machine. Several versions were offered and many sold in kit form until 1973 saw the last of the 3,300 Lotus-supplied cars. Nowadays Sevens are sold by Caterham (as shown) with the same fun value per mile and the same minimalist weather protection, ie none. Original Lotus Sevens are now worth a lot – £10,000 plus ($18,500) in excellent condition.*

Traditional classic cars suffered in this period too, since they became the focus of the investor market. Some attributed the tremendous rise in old car values to the death of Enzo Ferrari, sparking a groundless fear that no more true sportscars could ever be built. Others say that after the 1987 stock market crash around the world, financial investments were held off in favour of investments in tangible assets.

Whatever the case, values of perfectly usable Aston Martins and Ferraris suddenly shot up, tripling and quadrupling in a matter of months. Ironically, these same Ferarris had been impossible to move from American used-car lots, just a few years earlier, and 308s had reputedly sold for around $4,000 (£2,160).

This classic car fever resulted in the price for average cars in moderate condition rising so steeply, that they could stand horrendously expensive complete restoration to concours condition and still earn their owners a fat profit. Late 1990 saw a fall in the prices realized for such prestige classics, with many auctions failing to sell even 50 per cent of their lots.

So we come to the cars of tomorrow. The selection I made was difficult, not only from the point of view of selecting an even balance of dream cars and more regular usable cars, but also because when you do study the market from 1980, there are a great many vehicles from which to choose.

Leaving aside for a moment cars which are due out soon such as the new Bugatti, the McLaren ultimate road car and the road-going Porsche 956, there were cars such as the Dodge Viper, the Porsche 944 Turbo, the VW Corrado, the Ford Capri 2.8i, the Rover Vitesse, the Vauxhall/Opel Lotus Carlton/Omega and the Caterham Seven, all worthy of consideration and potential future investments.

To me – and I hope to you – the sixty or so cars that I have chosen will become classics, maybe not tomorrow, but in years to come. They all have some elements of style and elegance, and in most cases they are fine performers. Since the earliest motoring dawn, man has been fascinated by speed, and gripped by an overwhelming desire to

RIGHT
*Ferrari can boast one of the most impressive and prestigious ranges of supercars in the world, a range which includes the 308 GTS Quattrovalvole pictured here.*

go as fast as possible encased in a steel carriage. Now, with cars capable of incredible speeds, emphasis focusses on designed-in safety, in the shape of crumple zones, and anti-lock brakes.

Some of the cars I have included are dream cars, built in minute quantities, superb speed machines, dramatically styled and often immaculately finished. Others are more 'mundane' everyday cars in which you can pick up the kids from school, go shopping or crawl through city traffic, without engines protesting. These cars are as much future classics as the dream cars, since the urban sprawl of the future won't be any less congested than it is now. Fuel-efficient cars will be appreciated in twenty years' time as much as they are today.

Apart from speed and style, the majority of the cars are special production models, some are homologation specials, that is they were built to satisfy the laws of racing, whereby a manufacturer has to build a certain number of cars before he can let loose either identical, or highly modified versions on the race track. Good examples of these are the Ford RS Cosworth 500, the Evolution 11 Mercedes 190E 2.5-16 and the evergreen Audi Quattro, possibly one of the oldest cars in the book, but as popular today as ever and likely to become more collectable, now that production has been halted.

BELOW LEFT

*The Dino replacement, the 308 GT5, GT5i and GTB, sold in vast numbers in Ferrari terms; over 6,000 of all types, with fuel injection for the quad cam V8 after 1981. Prices rocketed in late 1980s, but are now more stable. It is undoubtedly a brilliant car with stunning looks, aggressive performance and sure-footed roadholding. Although a bit obvious, it is still a great buy at the right price – £45,000 ($83,000).*

The wonderful thing about these classics of the future is that they will have been designed with longevity and ease of maintenance in mind. The common classic, if you like, will be easier to maintain twenty years from now, and for most cars, lesser siblings will have been produced in such volumes that a body panel or mechanical component will be readily and, hopefully, inexpensively available from any number of outlets, so aiding running and restoration. The Ford Escort RS Turbo provides an excellent example.

Whichever cars you mark down for the future, just remember that while some of the cars in this book may make good financial investments, all the cars included were built by their designers to be driven. From Ford to Ferarri, nothing hurts a talented designer more than to see his offspring penned up in a dusty garage.

ABOVE

*The wolf in wolf's clothing; the Porsche 944 Turbo a superb touring, sporting, driving machine, with flared arches to distinguish between the 3.0-litre non-turbos and 2.5-litre turbo cars. The performance (250 bhp/6000 rpm and 258 lbft/4000 rpm) from its 2.5-litre four-cylinder engine is exciting with a top speed of 150 mph (241.5 kph) plus. Lots were made but it is still worth getting hold of one. Prices start from £20,000 ($37,000) for good second/third hand models.*

RIGHT
*One of Zagato's most inspired shapes, the DB4GTZ, was a track star in the early 1960s. Values today have soared, though they were never cheap to buy.*

If you do want to buy a car and keep it as pristine as possible, don't forget that standing still for weeks at a time does a car no good. Fuel pumps, cooling systems and oil ways need to be flushed through to stop them silting up, and the best way to do this is to warm up the engines as often as possible.

Treat a cold engine gently, and always run it until it reaches operating temperature, so that all the functions work properly. A quick ride around the block won't add hundreds of miles to the odometer, but it will keep the gearbox and drivetrain in peak condition.

Laying up for the winter requires special preparation and thought. If you're unsure how to tackle this, then magazines such as *Classic and Sportscar, Your Classic, Classic Cars* and *Practical Classics* give regular sound advice on the subject. Your Classic even has its own telephone helpline, which receives calls from all over the world daily.

If you're choosing a car as an investment here's a rough guide. Choose a car from a period when a manufacturer was doing well, say with a successful race team. Convertibles and coupes are the best styles to buy followed by saloons and station wagons.

If the car you buy has been restored, make sure the job was done by a reputable company. If you're paying out large sums, ensure that the car is as originally finished as possible, not modernized, customized or fitted with pirate panels. Buying original, unrestored cars is the wisest option, but check that all the bits are intact, as some may be unobtainable. Finally, don't forget to obtain a special classic car insurance policy with an agreed value. To insurance companies, some so-called classics are just old bangers.

But the most important thing is to get out there and enjoy the driving experience. It's what classic motoring is all about.

Alexander Corne
*Future Classics*

# ASTON MARTIN VIRAGE

The Virage is the first new volume production Aston for 20 years, succeeding the AM V8, an act which is one of the industry's hardest to follow. Volume production in Aston Martin terms is minute of course, and really means continuous. Designed by two talented British designers, Heffernan and Greenley, the sleek coupe looks a lot smaller than the thumping V8 it replaces. There's no lack of charm in the body close up though, and the now traditional AM grille survives, albeit in an amended form.

Launched with great aplomb at the 1988 NEC Motor Show in Birmingham, only minutes after Jaguar unveiled the XJ220, the Virage caught the imagination of the motoring world. It is designed for the 1990s, aerodynamic, and not a trace of a stick-on spoiler or other dubious aerodynamic aid in sight.

Inside there are the expected Aston Martin luxuries, wood and leather, the latest computerized information systems and traditional analogue instruments. Sink into the sumptuous seats and you'll be away in dreamland.

Demand of course exceeded Aston's supply lines even before the launch-day champagne bubbles had subsided, and waiting-list times are quoted in years, though Aston must be congratulated for not resting on these complacent laurels of full order books. The Volante convertible is now on the roads, after significant development work. The more powerful Vantage is straining at the leash, or so we're told.

The Virage provides stunning road-going performance, allied to a splendid and attractive interior: an all round success. As the values of the Newport Pagnell offspring often double as soon as they meet their new owners, it's unlikely many will be driven in the manner for which they were intended.

BELOW LEFT
*The Virage Volante arrived in 1990 to a long waiting-list and an initial price tag of £125,000 (c.$231,000).*

Traditional to the last, the Virage features the usual longitudinal engine configuration, a quad cam V8, and rear-wheel drive. New refinements include four valves per cylinder, a catalyst and fuel injection, and a new rear suspension, complete with cast trailing arms in light alloy and Watts linkage and an alloy De Dion tube.

The Virage has to take Aston Martin into the next century and so has been created using the latest in computer-aided design, as well as some of the brightest brains around. The chassis

## TECHNICAL SPECIFICATION

| | |
|---|---|
| **MODEL** Aston Martin Virage | **TOTAL PRODUCTION** 160 to date |
| **ENGINE** 5340cc, V8, 32 valves, DOHC per bank | **NUMBERS BUILT P/A** 300 (projected) |
| **POWER** 330 bhp @ 6000 rpm/340 lb/ft @ 3700 rpm | **ENGINE/CHASSIS NOS** N/A |
| **CHASSIS** Steel/aluminium bodyshell | **SPARES AVAILABILITY** All parts available |
| **BRAKES** Ventilated discs/drums, independent system front & rear | **PRICE AT LAUNCH** c.£120,000 (c.$222,000); $225,000 |
| **TOP SPEED** 155 mph (248 kph) | **OPTIONS** Colours: Customer specification<br>Trim: Customer specification<br>OE Options: Electric mirrors, windows, trip computer, leather interior, air conditioning |
| **ACCELERATION** 0–60 mph: 5.5 secs | |
| **PRODUCTION SPAN** 1990– | |

BELOW RIGHT
*The 1988 Virage Coupe's styling was markedly different from the old V8 it replaced. Despite the radical styling, lack-lustre performance earned it criticism.*

is stiffer and the body more aerodynamic, providing easier handling and better performance. The Virage won't be a brute to handle, and of course superb quality is built in by hand.

Owning a Virage will be a luxury only the very committed and wealthy will enjoy. With prices now in the motoring stratosphere, even dedicated museums will struggle to obtain one. With the arrival of the new stylish convertible in 1991, interest is sure to peak again. The Virage comes as a 'world car', designed to meet all motoring regulations. The Virage must be one of the best cars in the world, but the best *job* in the world must be selling them.

PAST CLASSICS

## ASTON MARTIN V8

The Aston Martin V8 began life in 1969 as the DBS V8 Saloon, complete with 325 bhp from the massive 5.3 V8 and a stylish two-door fastback body. By 1972 the car had revised front fittings giving only two headlamps, and the humped grille returned. Almost 2,000 of these V8s were produced by the time production gave way to the Virage, and although they were gas guzzlers supreme, the V8s had a loyal and monied group of devotees throughout the world. Performance packages were called Vantages and Volantes and sported soft tops. Vantage Volantes were the most coveted of all V8s produced in the last two decades, capable of extracting 170 mph (272 kph) from the 438 bhp engine. Find one and store it. It will be a gold mine in years to come.

RIGHT
*The AM V8 was a successful seller, seen here in basic guise, though the Vantage, a higher performance model, is more coveted. The drop-head Vantage Volante is one of the most highly prized Astons ever.*

# ASTON MARTIN ZAGATO COUPE/CONVERTIBLE

Aston Martin and Italian stylist Zagato got together to produce a dramatically styled coupe in 1986. The traditionalist car producer was making a bold statement, bringing in the Italians once again to liven up their staid-looking, muscular supercars. Following the immensely successful launch of the car to an adoring public, Zagato subsequently designed a softer-looking convertible version in 1988.

The aluminium body was strikingly angular and mated to a formidable steel chassis. Power came from the 5340cc V8 and was transmitted through either a three-speed automatic or five-speed manual box. Inside there were acres of soft leather, with walnut dash and cappings surrounding the instruments and electronic switchgear. Unlike many of its peers gimmickry was notably absent from the options list; there was no four-wheel drive or four-wheel steer, no superfluous trip computers, and no ABS. The rolling chassis was built in England and then shipped to Italy for the body to be fitted.

The two different body styles were both powered by the same V8 engine, but tuned differently. The Vantage, built between 1986 and 1988, had carburettor feed and produced 432 bhp, reached a top speed of 185 mph (296 kph) and managed 0–60 mph in 4.8 secs. It weighed in at 1650 kg.

The Volante, built between 1987 and 1989, had a more refined and less rushed air about it. The engine used a fuel-injection system, and was tuned to a relatively modest 305 bhp. There was nothing slouchy about the top speed of 165 mph (265 kph), nor the 0–60 mph time of 6.0 secs. It weighed in at 1,780 kg, what with the power hood's hydraulics and all. The Volante was tuned to a slower specification as it was felt more appropriate to the design and potential customer.

ABOVE AND BELOW
*The Zagato Coupe looked stunning, and with only 50 produced, it will always be sought after. Prices at auction have levelled off at around £200,000 (c.$370,000).*

Aimed at the classic collector and investor, these Astons were incredibly expensive, and also produced in minute numbers even by Aston standards. Starting off at £88,000 ($163,000) for the first coupes, by 1989 convertibles were retailing at £175,000 ($324,000). Demand was such that second-hand prices reached as much as £400,000 ($741,000) by 1990.

Zagatos have instant classic appeal as traditional, British grand tourers. Selling the cars to many celebrities also boosted the image, but the inherent strength of build, and precise, skilled engineering, makes any Aston a timeless classic. The Zagato is even more so, due to its radical looks and low build numbers.

The last eight to ten years have seen the values of Aston Martin cars rise without precedent as investors rushed to salt them away. Prices have levelled off, but the Zagato is still good value, as less than 100 of all types were built, and most have been looked after like rare oil paintings. Any Zagato under £200,000 ($370,000) should be viewed as a good purchase.

## TECHNICAL SPECIFICATION

| | | | |
|---|---|---|---|
| **MODEL** | Aston Martin Zagato Coupe/Convertible | **TOTAL PRODUCTION** | Vantage 50/Volante 35 |
| **ENGINE** | 5340cc, V8, DOHC per bank | **NUMBERS BUILT P/A** | c.15 |
| **POWER** | 432/305 bhp @ 6200 rpm/395 lb/ft @ 5100 rpm/ 320 lb/ft @ 4000 rpm | **ENGINE/CHASSIS NOS** | N/A |
| **CHASSIS** | Steel/aluminium body | **SPARES AVAILABILITY** | All parts available ex-factory |
| **BRAKES** | Discs all round | **PRICE AT LAUNCH** | £88,000 ($163,000); $175,000 |
| **TOP SPEED** | 185/165 mph (296/264 kph) | **OPTIONS** | Colours: Customer specification<br>Trim: Customer specification<br>OE Options: Air conditioning, leather interior, trip computers electric windows |
| **ACCELERATION** | 0–60 mph: 4.8/6.0 secs | | |
| **PRODUCTION SPAN** | 1986–1989 | | |

RIGHT
*A bold Zagato interior is not subtle, but functionally pleasing. Cars which are driven hard don't wear well. Most, however, have been preserved in excellent shape.*

PAST CLASSICS

## ASTON MARTIN DB4 GTZ

The last time Aston Martin and Zagato got together was in 1960 when the Italian design maestro produced a lovely shape to clothe the DB4 chassis. The aggressively styled DB4 GTZ had a lightweight body and was perfect for racing. Although the car was in 'production' between 1961 and 13, only 19 were ever made. The 3670cc double-overhead-camshaft, six-cylinder engine pushed out 314 bhp and reached 60 mph in 6.9 secs. These cars are extremely rare and worth approximately £750,000 ($1.38m). To the purist the DB4 GTZ will be more desirable than the 1980s car, since it had a competition history and was a fully developed derivative of a great production car. The later Zagato was designed to attract attention in the street, not to win its spurs on the track.

RIGHT
*One of Zagato's most inspired shapes, the DB4GTZ, was a track star in the early 1960s. Values today have soared, though they were never cheap to buy.*

# BENTLEY TURBO R

Don't be fooled by appearances. This is not a Rolls Royce dressed up for affluent street racers. Bentleys cat-napped in the shade of the more glamorous Rolls Royces for over 50 years while in the hands of the engineers from Crewe, but reassuringly, the name never died. Come 1982 and the Bentley name had a respectable rebirth on the back of a car designed to revive the spirit of times past. Recounting the glories of the 'Blower' days, the car was named Mulsanne, after the famous straight at the Le Mans circuit, which saw some of the most famous Bentley achievements of all.

Since then Rolls Royce has developed the marque so that it now no longer means Rolls Royce by another trim level, but represents a more serious sporting approach to wielding over two tons of the finest handcrafted, motorized luxury money and status can buy.

Deep down inside the turbocharged engine lies the indomitable V8 6.75-litre unit, the power output of which Rolls Royce haughtily refuses to divulge. A less expensive Bentley 8 was offered from the mid-1980s and this proved to be a popular option. Sales of the Turbos caught on too, and by 1990, the Bentley marque was providing the mass seller Rolls Royce always desired.

Not wishing to detract from the cachet of owning a rare car, Rolls Royce nevertheless wanted to increase their sales. Selling several thousand 'Bentleys' every year and far fewer Rolls Royces, meant better annual turnover, while retaining the exclusivity of the Rolls Royce name.

The Turbo R is a fast motorcar; it will pull to almost 140 mph (225 kph) while never stemming an unquenchable thirst for fuel. The silent express now comes with self-levelling suspension for an even smoother ride at speed, though it could never be accused of being overtly sporty.

Finished to Rolls Royce standard, the Turbo R has timeless elegance and breeding, setting it apart from those pretenders, keen to snatch away the prestigious and much-coveted title, 'the best car in the world'.

The Turbo R will always be a sought-after car, since it offers true engineering excellence, a high degree of status value and a performance package which is remarkable. Owning one will never be cheap, and restoring one will be a bank manager's nightmare. A pristine Turbo R will be a car to pass on to your grandchildren as a memento of an age when fuel consumption and running costs were, to the privileged few, an irksome irrelevance.

BELOW LEFT

*The fastest road-going Bentley for many, many years, the Turbo R is an expensive but rewarding carriage, well engineered and exquisitely crafted.*

BELOW RIGHT
*Obliged by the dire financial state of his company, WO Bentley dropped his objection to supercharging and allocated a corner of the Cricklewood factory to producing touring sportscars with a blown engine. The result was the 1930 4.5-litre 'Blower' Bentley. Despite weighing in at over two tons, the 26 open Vanden Plas-bodied models at least looked as though they could grace any race track.*

The Bentley 8 is a sensible starter classic for the aspiring Rolls Royce collector. With its distinctive and evocative mesh grille, the performance from its V8 is still more than acceptable, and interior luxuries adequately set it apart from other highly specified super saloons.

Service histories from respected dealers are obviously vital but an enterprising buyer might consider a trip to the USA or the Middle East to find a well-cared for example, although several have been outrageously tarted up by monied and tasteless owners.

## TECHNICAL SPECIFICATION

| | |
|---|---|
| **MODEL** Bentley Turbo R | **TOTAL PRODUCTION** N/A |
| **ENGINE** 6750cc, V8, Electronic fuel injection | **NUMBERS BUILT P/A** N/A |
| **POWER** 333 bhp @ 4500 rpm/486 lb/ft @ 2250 rpm (unoff. figs) | **ENGINE/CHASSIS NOS** N/A |
| **CHASSIS** Unitary steel | **SPARES AVAILABILITY** All parts available |
| **BRAKES** Discs/discs, ABS | **PRICE AT LAUNCH** c.£68,500 (c.$127,000); $174,000 |
| **TOP SPEED** 145 mph (232 kph) | **OPTIONS** Colours: Customer specification<br>Trim: Customer specification<br>OE Options: Air conditioning, leather seats |
| **ACCELERATION** 0–60 mph: 6.5 secs | |
| **PRODUCTION SPAN** 1985 | |

PAST CLASSICS

## 1930 BLOWER BENTLEY 4.5 LITRE

A hat-trick of victories at Le Mans during the last years of the 1920s made Bentleys the most famous racing cars in Europe, and from the development of the 3.5-litre engined cars came the 4.5-litre models. With the help of Amherst Villiers, the supercharged 4.5-litre cars later ruled the race tracks and created the legend best encapsulated in the term British Racing Green, the colours of the Bentley Racers. The Bentley Boys were some of the most famous men in pre-war racing: Captain Tim Birkin, Woolf Barnato, Glen Kidston, Sammy Davis and Dr Benjafield. Today their cars and others from the same era command vast prices, but for those without a budget stretching into millions of pounds or dollars, the Bentley Turbo R is as close as you'll get.

# FORD ESCORT RS TURBO

Ford has captured the go-faster market with hot Escorts since the late 1960s. The current fashion-conscious buyer-propelling carriage is the RS Turbo, complete with quite restrained body-kit and nondescript paintwork. Outwardly the RS Turbo doesn't scream 'highly charged, take care!', but on the road few people get to study more than the rear panel.

At rest it is probably the most visually appealing Escort ever made, save for the Cabriolet model, which has an element of style, sadly lacking in much of today's range of blue oval-badged cars.

Under the bonnet Garrett Research's T3 turbo sends the pocket road-burner up to 60 mph in 8.1 secs and leaves it breathless at over 130 mph (209 kph). A limited slip differential with viscous coupling enables the RS Turbo to corner as if the tyres were coated in super-glue. High cornering speeds simply encourage the car to stick even tighter; the car only becomes unsettled if you lift off in mid-bend.

Suspension follows the RS 1600i formula of longitudinal front-wheel locating bar, MacPherson struts and anti-roll bar at the front, and coil springs and Girling dampers at the back, like the XR3i, but with an anti-roll bar there too.

The RS 1600i incidentally, is another limited edition, motorsport-inspired Escort you might like to look into. Only a small number were built, with loads of motorsport running-gear modifications. They're identifiable by their graded stripes across the bonnet, and RS 1600i decal quietly stuck across the rear hatch.

So the RS Turbo currently languishes with the street rod crowd, failing to capitalize on the image of the RS Cosworth Sierras which, being strictly fair, have the same built-in deprecating factor of being, underneath it all, just tarted-up, mass-produced Fords.

But see the Escort RS Turbo as just another Escort and you're missing the point, which is that the car is remarkably refined for all its neck-bending acceleration and top-end performance. For the money though, there's more panache and style on offer in other showrooms.

The Sierra succeeds above the Escort in the search for status because it is a more refined and better-known racer on the tracks. It also carries the Cosworth name; stick that on the Escort's notchy rump and see how sales take off. The RS these days is lost without the magic of Cosworth.

BELOW AND BOTTOM
*Despite being one of the fastest Fords on the road, the Escort RS Turbo lacks the Cosworth name and the associated status of the RS Cosworth Sierras.*

## TECHNICAL SPECIFICATION

| | |
|---|---|
| **MODEL** Ford Escort RS Turbo | **TOTAL PRODUCTION** MkI (Diamond White, codenamed White Lightning) 8,000; MkII (all colours) 20,000 plus 1986 and 1990 |
| **ENGINE** 1598cc, 4 cyl, Garrett T3 Turbo | **NUMBERS BUILT P/A** c.4,500 |
| **POWER** 132 bhp @ 5750 rpm/133lb/ft @ 2750 rpm | **ENGINE/CHASSIS NOS** N/A |
| **CHASSIS** Unitary steel | **SPARES AVAILABILITY** All parts available |
| **BRAKES** Ventilated discs/drums, ABS | **PRICE AT LAUNCH** £9,951 (1984) (c.$18,430); not officially exported to North America |
| **TOP SPEED** 128 mph (205 kph) | |
| **ACCELERATION** 0–60 mph: 8.1 secs | **OPTIONS** Colours: White, red, black, grey. Trim: Zolda, velour in shadow colourway. OE Options: Fuel computer, custom RS pack included Recaro seats, central locking, electric windows |
| **PRODUCTION SPAN** 1984–1990 | |

For future reference, the RS Turbo will not be a difficult car to buy. Spares will never be a serious worry, since so many bread-and-butter shells and running gear units have been produced. Garrett will be able to supply turbo bits at a price and this is really the only area of concern. New turbos don't come cheap and in all probability will require replacement at least once. Maintaining the correct oil for the turbo is of paramount importance, as is regular changing of this vital lifeblood. Check service records closely.

Fast Fords are fun, but buy with prudence; many will have had hard lives, so don't expect a pristine car for little money. If you take the hunt for a collectable RS Turbo seriously, then the talk is that the earlier Series One cars are the ones to buy. The Custom RS Turbo has most appeal, with a smarter, sportier interior.

PAST CLASSICS

## FORD ESCORT RS 2000

The RS 2000 was Ford's way of converting rally wins all through the 1970s into road-going sales successes. The RS 2000 was a highly specified street machine, with true rally-bred credentials. Available both in MkI and MkII shells, the later cars were discernible by the plastic, slanting nosecone, carrying four round headlamps. Earlier RS 2000s had garish body stripes and wacky 1970s paint colours, including deep purple and bright orange. Most have had race or rally experience, and few except true one-owner enthusiast cars can be trusted to be straight. Few are on the roads now, most are in safe keeping. Power from the MkI 1993cc overhead-camshaft engine was 100 bhp and 110 mph (177 kph); the later MkII had 110 bhp and 110 mph (177 kph), and was available with Custom and Custom X packs with more goodies and bodykit extensions.

RIGHT
*Earlier RS 2000 MkI-shape cars are now worth far more than later 1970s MkIIs. Here is the MkII RS 2000 alongside the slightly more aggressively styled Custom Pack model (green car). Note the popular 1970s vinyl roof option.*

# FORD SIERRA RS500 COSWORTH

Designed as a homologation special for Ford's serious competitions department, the RS500 was tunable to 400 bhp by changing the engine management systems' chip. Road-use cars weren't as tractable, but on the circuits the race cars cleaned up impressively. They were definitely the cars of their generation, and fiercely chased by the likes of BMW. The RS500 has superb historic racing potential, and road cars are great classic buys, prices though will always be at a premium, so buy whenever you can.

Differences which set the RS500 apart from the 5000 original-spec Cosworths are a fatter spoiler at the front, and twin spoilers at the rear, the upper unit having a thicker upturned lip. An RS500 decal adorns both flanks forward of the doors. Dunlop D40s adorn the alloy wheels and, to keep the larger brakes from boiling, there are air-cooling ducts built into the front bumper assembly.

Mechanically the RS500 was way ahead of the original car, having a far pokier motor. There were Mahle pistons, two fuel injectors per cylinder and thicker walls to the iron cylinder block. Suspension settings remained unchanged. Downforce was increased dramatically by the new spoilers and, to haul in this three door supercar, the TEVES anti-lock brake system helped the four-wheel ventilated discs cope effectively.

Built not by Ford nor its Special Vehicles Operations department but by Aston Martin Tickford, all 500 cars were turned out in a record time of seven months. The deal with Tickford came about after they had built the Tickford Capri for Ford between 1982 and 1987. Ford bosses were never entirely happy with the be-spoilered Capri 2.8T, but the RS500 was far more happily welcomed into the Ford fold. The car had dynamite racing potential and, as it turned out, the RS500 was exactly the track terror Ford hoped it would be.

ABOVE

*It's a tight fit squeezing all those 400 horses under the broad Sierra bonnet. Performance was superb from the twin cam turbocharged Cosworth engine.*

The RS500 was a collector's car immediately it was born. Despite the limited numbers, it was the combination of Ford reliability and engineering along with the Cosworth motor and Tickford coachwork which drew in the crowds. Racegoers marvelled at track success after track success, and even now, three years after their debut, the track cars are still pounding the circuits and collecting the silverwear.

The car was in such demand by the racing fraternity that RS500 body shells returned to production in 1989, though they were for to race-use only. A road-going car won't drop in value much below its purchase price. Its lesser brother the 'standard' RS Cosworth was fetching good sums even before the body shell change to four doors in 1987/88. Expect to pay a large sum for your RS500, but once you've laid your hands on one, you'll have a classic Ford to cherish.

## TECHNICAL SPECIFICATION

| | |
|---|---|
| **MODEL** Ford Sierra RS500 Cosworth | **TOTAL PRODUCTION** 500: 393 Black; 52 Moonstone Blue; 52 Diamond White |
| **ENGINE** 1994cc, turbo, 16 valves, DOHC | **NUMBERS BUILT P/A** 500 |
| **POWER** 224 bhp @ 6000 rpm/206lb/ft @ 4500 rpm (road spec) | **ENGINE/CHASSIS NOS** GBBE GG 38600–39099 (chassis ID); eng nos. YBD 0015–0537 |
| **CHASSIS** Unitary steel | **SPARES AVAILABILITY** Pretty good, body shell back in production |
| **BRAKES** Discs all round, ABS | **PRICE AT LAUNCH** £19,950 (c.$36,907) |
| **TOP SPEED** 150 mph+ (240 kph+) | **OPTIONS** Colours: Black, white, metallic blue<br>Trim: Leather<br>OE Options: Higher performance engine technology management system |
| **ACCELERATION** 0–60 mph: 6.1 secs | |
| **PRODUCTION SPAN** 1987 | |

ABOVE
*The Sierra RS500 Cosworth, with its 400 bhp engine, is a higher development of a successful theme. Cleaning up on the tracks still, the road cars are gems to preserve. Tuned 400 bhp engines are wild fun.*

PAST CLASSICS

## LOTUS CORTINA MK1 1963–66

Racing hotted Ford saloons isn't a new idea. Back in the early days of the Cortina's life, Lotus and Ford conspired to dominate the race tracks with this 110 mph (177 kph) twin cam gem. Easily identifiable by the cream paint and green Lotus side stripe, wider though plainly hub-capped wheels and those chubby quarter bumpers, the Lotus Cortina cleaned up on the circuits, piloted by some of the fastest men of the day.

For the road, the 1600cc cars proved to be highly successful, and good examples now fetch around £15,000 ($28,000). Don't be misled by the later boxy Cortina Lotus, a Ford-built MkII Cortina with the Lotus goodies plumbed in, the real McCoy comes only in MkI form, despite later MkIIs being highly effective racers too. Just over 4,000 MkIs were built in the three-year production life, with similar numbers of MkIIs later till 1970.

RIGHT
*Race-track successes boosted both initial sales and later collectability of the Lotus and Ford progeny. Between 1963 and 1966 over 4,000 Lotus Cortinas were produced.*

# JAGUAR XJ-S CONVERTIBLE

The longest-awaited Jaguar ever, the Convertible XJ-S, appeared more than ten years after the XJ-S coupe was launched to a critical world in 1975. This rag-top is a refined Jaguar, stylishly converted, losing none of the XJ-S's appeal, which has grown with each passing year.

Propelled by the thirsty alloy 5345cc V12, the Convertible has a power hood, which is fully lined and stuffed, complete with a glass rear window. The loss of the coupes 2+2 status precludes the need for the cars to meet stringent four-seater safety and crash legislation.

The Convertible body shell took three years to perfect, having strengthened 'A' posts, sills and underfloor. Bulkheads were reinforced, but the crumple zones, which protect life and limb during accidents, are the same as the coupe's, which is an indication of the inherent solidity of the car's design.

The Convertible is linked to the playboys of the 1990s in the same way that the Rolls Royce Corniche was tied to property developers in the early 1970s. It combines style and panache with grand touring abilities, which are hard to beat.

The body may flex a little under the hard forces of enthusiastic driving or the failings of badly made roads, the ride might be a little too soft, but the V12 loves to be stretched and, at high speed, passengers aren't buffeted from side-to-side. In-town docility and the lack of any mechanical hystrionics make this Jaguar simple to maintain. Regular servicing by the Jaguar agent is, of course, a necessity but need only be carried out at lengthy intervals.

The XJ-S Convertible launch saw the introduction of Marelli digital ignition, as well as the fitting of the TEVES anti-lock braking system as standard across the V12 Jaguar range. Demand has held up since day one, despite a greater than 25 per cent price hike in two years.

XJ-Ss can rust though, and the Convertibles will be no different. If you look at a car which hasn't been maintained and has lived in a damp climate, check the underside carefully. Jaguar V12 engines will also be quite pricey to repair, so make sure the car you set your heart on has a full service history from an approved dealer. Remember alloy engines need anti-freeze all year round.

BELOW

*Some now say this is Jaguar's best-looking convertible yet. Over a decade after the much-maligned Coupe appeared, the drop-head won many hearts and sales even with prices at around £40,000 ($75,000). The usual V12 power, gracious looks and relative rarity mark this car down for future greatness, and demand is still strong so prices will remain firm.*

## TECHNICAL SPECIFICATION

| | |
|---|---|
| **MODEL** Jaguar XJ-S V12 Convertible | **TOTAL PRODUCTION** N/A |
| **ENGINE** 5345cc, V12, electronic fuel injection | **NUMBERS BUILT P/A** c.5,000 |
| **POWER** 286 bhp @ 5150 rpm/310lb/ft @ 2800 rpm | **ENGINE/CHASSIS NOS** N/A |
| **CHASSIS** Steel unitary/extra chassis stiffening | **SPARES AVAILABILITY** All parts available |
| **BRAKES** Ventilated discs/discs, ABS | **PRICE AT LAUNCH** £36,000 ($66,600); $56,000 |
| **TOP SPEED** 145 mph (232 kph) | **OPTIONS** Colours: Standard Jaguar solid and metallic at no extra cost Trim: Herringbone tweed cloth or leather interior at no extra cost OE Options: Catalyst exhaust, sports suspension, electric sunroof, telephone, seat piping |
| **ACCELERATION** 0–60 mph: 8.0 secs | |
| **PRODUCTION SPAN** 1988– | |

## PAST CLASSICS

# JAGUAR E-TYPE

Jaguar's E-Type was possibly the most charismatic car of the 1960s. Stunning audiences at its launch in 1961, the svelte lines and race-proven engineering helped sales to soar. Over 77,000 of all types were sold between 1961 and 1975 when the last V12s died. Sold as a convertible, 2+2 roadster, two-seater coupe and 2+2-seater coupe, the E-type is as popular now as any time during production. V12 Es are identified by the cross-bar grille and special undertray, which sucked air for four Zenith Stromberg carburettors. Prices are now descending after riotous behaviour during the late 1980s. The E-Type rusts badly if neglected and the V12 needs constant tuning and nursing. Top speed is close on 150 mph (242 kph) though emissions controls reduced that by 10 to 15 per cent.

RIGHT
*The E-Type took the D-Type concept and adapted it for the road, making the latter's level of performance available to the man on the street. Open and closed body styles were offered.*

# JAGUAR XJ220

The indisputable star of the British Motor Show in 1988, the Jaguar XJ220, captured the imagination of the motoring public in a similar way to the E-Type in 1961 and the XK120 in 1948. Jaguar had done it again.

The XJ220 was originally a styling exercise, built up by a small team of the company's highly skilled engineers, who used lots of their own time to develop a few scratchy ideas into the most stunning shape to emerge from the Coventry manufacturer for many, many years.

The original show car was to be powered by a development of the V12, bored out to 6.2 litres, equipped with 48 valves, double overhead camshafts per bank and tuned up to 500 bhp. Maximum speed was predicted to be 200 mph+ (320 kph) and 0–100 mph was quoted at 8.0 secs. Four-wheel drive was standard.

The unitary frame or 'tub' was constructed from bonded aluminium, with a steel roll cage for extra safety. The car could almost be considered as a road-going version of the company's successful World Sports Prototype Championship race cars, which that year had scooped Le Mans for the first time in 30 years.

Jaguar were literally overwhelmed by the response at the Motor Show, £50,000 ($92,500) deposit cheques for production cars which weren't even planned, were waved furiously. Only after the show, when the impact of the XJ220 had sunk in, did they seriously consider full-scale production.

Now, as we enter the 1990s, the slightly re-revised XJ220 is about to enter production. Jaguar has decided that 350 will be built, albeit on a wheelbase 10 inches shorter and now powered by a version of their other successful racing engine, the V6 3.5-litre. Four-wheel drive has been dropped in favour of weight-saving two-wheel, rear-wheel drive.

ABOVE
*The super luxury of the sports-orientated XJ220's interior. All cars will eventually be trimmed to personal tastes, although the factory design is quite stunning.*

Jaguar has the jump on the other WSPC racing manufacturers, aiming to have their race-derived road car ready well before Maclaren get their 'Ultimate Road Car' off the blocks, and also ahead of Porsche's much vaunted road development of their 956. What this means is that the car will never sell for less than a quarter of a million pounds (c.$460,000), demand is likely to remain strong for decades, and few will turn a wheel in anger on the roads where they can be involved with side swiping saloon drivers and suicidal juggernauts.

The cars are to be built by JaguarSport, based in Kidlington, Oxford. Deliveries aren't planned until 1992, and production is scheduled at roughly four cars per week. Jaguar relieved prospective owners of the £50,000 ($92,000) deposit cheques during early 1990, and a large queue has built up for the cars.

Although most XJ220s may languish in dusty museums and private collections, when they do come out, be sure to go to watch, since they are some of the world's finest sportscars.

## TECHNICAL SPECIFICATION

| | |
|---|---|
| **MODEL** Jaguar XJ 220 | **TOTAL PRODUCTION** 350 |
| **ENGINE** 3498cc, V6 twin turbo, 24 valve quad camshafts | **NUMBERS BUILT P/A** 4 cars per week |
| **POWER** 500 bhp @ 6000 rpm/472lb/ft @ 5000 rpm | **ENGINE/CHASSIS NOS** N/A |
| **CHASSIS** Bonded aluminium tub | **SPARES AVAILABILITY** See Jaguars for details |
| **BRAKES** Ventilated four-pot caliper discs all round | **PRICE AT LAUNCH** c.£290,000 ($537,000) before taxes, not officially exported to North America |
| **TOP SPEED** 200 mph+ (320 kph+) | |
| **ACCELERATION** 0–60 mph: 4.0 secs | **OPTIONS** Colours: Customer specification Trim: Customer specification. O E Options: Leather interior, laminated glass all round, butterfly doors |
| **PRODUCTION SPAN** 1992– | |

ABOVE

*Jaguar stole the 1988 British Motor Show with the XJ220, insisting it would not reach production. Now a V6-engined, modified car is being readied for sale at around £300,000 ($555,000). North Americans won't be so lucky as the XJ220 won't be sold officially in North America.*

PAST CLASSICS

# JAGUAR XJ13

Never raced in anger, this V12 development of the 'C', 'D' and 'E' Type racers was written off quite comprehensively in 1967 during a shake-down test for the cameras at the British motor industry's test track.

The monocoque shell had the V12 5-litre engine mid-mounted and mated to a ZF gearbox. While it had only a brief active life in the hands of the Jaguar test drivers, it proved that the V12 engine would provide bags of torque and high terminal speeds, at one stage reaching over 160 mph (257.6 kph).

Once rebuilt after the massive accident, during which the car rolled and turned end over end, the XJ13 was consigned to museum status. It wasn't until the efforts of American racer Tom Walkinshaw paid off in the early 1980s that Jaguar was again victorious on the race track. Since then, two Le Mans victories have crowned the Coventry company in glory.

The mid-mounted 5-litre fuel-injected engine was a forerunner of the 5.3-litre V12 which continues beneath the bonnets of the XJ-S and V12 saloon.

RIGHT

*The XJ13 one-off sports racing car never saw track competition. High costs killed off the project and the Haynes and Sayer masterpiece was confined to languishing in museums.*

# JAGUAR V12 SALOON

Since its inception in the pre-oil crisis days of the very early 1970s, the V12 Jaguar engine has been revered as one of the most perfect engines of all time. Developed for the XJ saloon and shoehorned under the bonnet of the legendary E-Type, the 5343cc engine has always provided turbine-like smoothness in its delivery of power.

The XJ-V12 has remained in limited production since it has such a strong following throughout the world, and so brings in much prestige for Jaguar. Additionally, it is such a fine power unit that it would be cruel to kill it off wantonly. Despite a frightening thirst which Jaguar thought would preclude its sale in the US in the 1990s, and deterred them from planning for it to be fitted into their XJ40 models, the V12 is long-lasting and reliable, provided that servicing is regularly and competently carried out. Maintaining anti-freeze levels to prevent corrosion of the aluminium alloy engine is vital, for instance.

The old XJ body shell continues to sell after 20 years, albeit in MkIII form, but the earlier MkI body styles are becoming collectable. The purists say that the Daimler Vanden Plas Double Six, mechanically identical differing only in interior and exterior trim fittings, is the most desirable V12, since less than 350 were made in MkI style.

The V12 will sweep its occupants to a conservative 140 mph (225 kph) in silent splendour. Only above 3000 rpm can you discern the V12 busily spinning ahead of the cabin, and then the

BELOW
*The V12 saloon is Jaguar's longest-running model at present. Twenty years in production and still few can match the smoothness of the V12 powerhouse. Cheap to buy second-hand, but repair bill can be horrendous.*

**TECHNICAL SPECIFICATION**

| | |
|---|---|
| **MODEL** Jaguar XJ-V12 Saloon | **TOTAL PRODUCTION** N/A |
| **ENGINE** 5343cc, V12, electronic fuel injection | **NUMBERS BUILT P/A** c.2,000 (inc Daimler V12) |
| **POWER** 264 bhp @ 5250 rpm/283 lb/ft @ 3000 rpm | **ENGINE/CHASSIS NOS** N/A |
| **CHASSIS** Unitary steel | **SPARES AVAILABILITY** All parts available |
| **BRAKES** Ventilated discs/discs | **PRICE AT LAUNCH** £3,725 (1972); $11,025 (1973) |
| **TOP SPEED** 140 mph (224 kph) | **OPTIONS** Colours: Full range of standard Jaguar metallic and solid colours. Trim: Most leather equipped, several finished in cloth. OE Options: Air conditioning, electric windows and mirrors, sunroof, alloy 'Pepperport' wheels, limited slip differential, cruise control, trip computer (Catalyst exhaust at extra cost) |
| **ACCELERATION** 0–60 mph: 8.7 secs | |
| **PRODUCTION SPAN** 1972– | |

sound is a pleasing one. Motorway travel is deceptively fast and on most demanding roads the much copied all round independent suspension provides a steady ride. Anti-roll bars and anti-dive geometry save the car from undue body roll, though any combination of a wet surface and speed through bends is not a happy one.

Aimed at the successful businessman who doesn't need to worry about fuel consumption or servicing costs, the V12 is soon at home in the most prestigious of driveways. A superlative high-speed, long-distance cruiser, the car comes with a host of features as standard to keep its drivers happy. Air conditioning, leather interior, electric windows, power steering, disc brakes and limited slip differential, as well as cruise control are all V12 trademarks, and these days a catalytic converter is offered for £500 ($926).

The V12's most amazing facet is its price. When first launched in 1972 it was not expensive, and even today priced in the mid £30,000s ($50,000) the car is still inexpensive by comparison with the offerings from other European luxury car makers.

RIGHT
*The V12 saloon's interior was designed for captains of industry who appreciated craftsmanship and quality. Fully loaded with extras as standard, most run through automatic transmissions.*

On the debit side, during the 1970s Jaguar were plagued with unreliability problems, crass manufacture and little or no development. Sir John Egan turned Jaguar round, and today the V12 has to be one of the most outstanding saloons in the world, at any price.

PAST CLASSICS

## JAGUAR XJ-C

In the mid-1970s, while the world awaited the E-Type's successor, the XJ-S, Jaguar shortened the XJ6/12 chassis and chopped out the rear doors to create the XJ-C, a coupe so beautifully proportioned it could only have been a Jaguar.

Many large saloons have smaller, pricier, more prestigious siblings, but none matched the XJ-C on looks. With pillarless windows and often fitted with a vinyl roof, the coupes won many admirers, and although they became great rust traps, the V12 versions, of which 399 Daimler and 1,873 Jaguars were made, have become collectors' items. Even the more modest six-cylinder coupes, fitted with 4.2-litre engines fetch £10,000 ($18,500) in tidy shape, and good V12s will appreciate rapidly as time goes on. XJ-Cs were built between 1975 and 1977, and only just over 10,000 of all types were built.

RIGHT
*The Jaguar XJ-C is a fine, pillarless coupe, now fetching at least £10,000 ($18,500) in V12-engine form.*

# JENSEN INTERCEPTOR SERIES FOUR

The Jensen Interceptor began life in 1966 and was immediately hailed as a car ahead of its time. It combined smooth, powerful performance from a large-block American V8-engine with the svelte lines of a Touring-of-Milan styled body. It was crafted by hand from steel by the Italian coachbuilder Vignale.

In 1976, Jensen Cars called in the official receiver but the spares concern soldiered on alone. Many people tried to re-establish Jensen and recommence production of the very fast but incredibly thirsty cars. After several years building only a couple of Interceptors per year, production is in 1990 firmly back into double figures.

The Series Four car is hand-built as a four-seater saloon or a slightly more expensive, and just as impressive-looking cabriolet. The 1990 cars run on 5900cc V8 Chrysler engines and have automatic gearboxes. There are no new ground-breaking FF (four-wheel drive) cars, but earlier Interceptors are still around.

The Interceptor was killed off by the 1973/4 oil crisis. Now with the smaller engine and demand for a high-performance luxury car outweighing the penalties of colossal fuel consumption, the Interceptor rides again. The Series Four can manage 15 mpg (24 kpg) if driven carefully.

The cars are all finished to personal specification and include leather seating, electric front seats and a fire extinguisher. The convertible body has lost none of its charm and can compete effectively with cars such as the XJ-S convertible, despite costing twice as much as the Jaguar.

Apart from the stylish body and powerful engine, the Jensen rides and handles like a pedigree Grand Tourer. It tops out at 140 mph (225 kph) and has phenomenal mid-range acceleration, less than 5 seconds between 50–70 mph. Being built in such small numbers and costing almost £100,000 ($185,000), the car will certainly be a collector's item in the future. The bodies take eight weeks to hand-build, though the running gear is constructed from bought-in standard items, mostly American.

In 1990 there was a six-month delivery waiting list, the result of a deliberate policy of building the cars slowly to keep numbers down and rarity appeal up. Prices are set to remain stable as long as the company stays afloat and there is the promise of a spares back-up facility.

Older Series One-Three cars are increasing in value rapidly and good cars should fetch around £35,000–£40,000 ($65,000–$75,000) soon.

The earlier cars, built in the 1960s and 1970s had a shaky time until just a few years ago. Many parts were unobtainable and running costs extremely high. Even the rare and beautiful convertible's prices were low.

BELOW

*For ultimate rarity the Series Four Interceptors must rate highly. Barely a dozen are hand-built each year, and the convertible is even rarer. The list price is well over £120,000 (c.$222,000).*

## TECHNICAL SPECIFICATION

| | | | |
|---|---|---|---|
| **MODEL** | Jensen Interceptor Series Four | **TOTAL PRODUCTION** | c.5,000 (all types) |
| **ENGINE** | 5898cc, V8, OHV | **NUMBERS BUILT P/A** | 25–30 planned for 1990 |
| **POWER** | 247 bhp @ 4500 rpm/310 lb/ft @ 3200 rpm | **ENGINE/CHASSIS NOS** | N/A |
| **CHASSIS** | Tubular steel/steel body | **SPARES AVAILABILITY** | Series Four cars easy/older types problematic trim and body parts sourcing |
| **BRAKES** | Ventilated discs all round | | |
| **TOP SPEED** | 140 mph (224 kph) | **PRICE AT LAUNCH** | c.£100,000 (c.$185,000); not officially exported to North America. |
| **ACCELERATION** | 0–60 mph: 7.2 secs | **OPTIONS** | Colours: Customer spec. Trim: Customer spec. OE Options: Air cond. leather interiors, electric windows and mirrors |
| **PRODUCTION SPAN** | (Series 4) 1987– | | |

RIGHT
*The new drop-head Jensen Series Four is still popular in Europe, even though the selling price is now well into six figures.*

PAST CLASSICS

## JENSEN INTERCEPTOR FF

The earlier series Interceptors were notable for their vast American V8 engines and their incredible thirst. They truly were the dinosaurs put down by the oil crisis. However, the FF four-wheel drive variants were the technical whizz-kids of their age; it took other executive performance manufacturers decades to catch on to four-wheel drive as a marketing, and eminently useful, option.

Only 124 FFs were ever built into long-wheelbase MkII and MkIII shells between 1969 and 1971. Identifying marks are the double slats on the flanks behind the front wheelarches. The FFs used Dunlop-developed anti-lock brakes.

RIGHT
*Four-wheel drive Jensens were ahead of their time in the late 1960s, but today the FF system is much copied. The older Jensens are rare, valuable and expensive, but practical supercars, even if super thirsty.*

# LOTUS ELAN TURBO SE

Do I have to tell you that the new Lotus Elan has a hard act to follow? Is there any true blue Lotus fan out there without a pretty little Colin Chapman-inspired 1960s Elan tucked up tight in a warm garage? No, I guess not, but while Chapman is sadly no longer with us, the new Elan fortunately is.

Shaped by modern-day styling, that is with a broad steeply raked screen, drooping nose and stubby tail and powered by a Japanese-derived engine, driving the front wheels, the new Elan shares only name, backbone chassis design and driving experience with its long-dead sibling.

That the new Elan Turbo SE charges to 60 mph in 6.7 secs, or that it will race to 137 mph (219 kph) probably won't make all that much difference to the majority of buyers, who'll be trading in their downbeat convertibles and other old Elan substitutes faster than Lotus can count the coinage.

The Elan has style and carries on the old Elan tradition of nimbleness, surefooted handling and exciting driving, and with the top down, engine singing and wind rustling through your hair, you'll be back in time.

The Elan of the 1990s has a 1588cc Isuzu-sourced motor, complete with twin camshafts and turbo, anti-torque steer front suspension and a fibreglass body, which is remarkably shake-free. What the new Elan offers is fantastic ride and handling, as well as being one of the widest sports cars going. It trounces the opposition in the supersports drophead market on price, costing around £20,000 ($37,000) in the UK.

The Elan will be as popular as ever if past lessons on build quality have been learnt, and production doesn't lead to bottlenecks and therefore black market-inflated prices. Definitely the most promising sportscar to come out of a British assembly plant in a long time; the ghost of traditional British convertibles still lingers over America. But wait till they get their hands on these latest cars, and see the opposition run. The real British sportscar is back.

## TECHNICAL SPECIFICATION

| | |
|---|---|
| **MODEL** Lotus Elan Turbo SE | **TOTAL PRODUCTION** N/A |
| **ENGINE** 1588cc, 4 cyl, 16-valve DOHC, fuel injection | **NUMBERS BUILT P/A** c.3,000–4,000 |
| **POWER** 165 bhp @ 6600 rpm/148 lb/ft @ 4200 rpm | **ENGINE/CHASSIS NOS** N/A |
| **CHASSIS** Steel backbone/fibreglass body | **SPARES AVAILABILITY** All parts available |
| **BRAKES** Ventilated discs/discs | **PRICE AT LAUNCH** c.£20,000 (c.$37,000); $39,040 |
| **TOP SPEED** 137 mph (219 kph) | **OPTIONS** Colours: Standard range of colours Trim: Leather complete with red stripe across seat backs |
| **ACCELERATION** 0–60 mph: 6.7 secs | |
| **PRODUCTION SPAN** 1990– | |

LEFT
*The new Elan has won many a purist's heart despite the Japanese origins of its engine and front-wheel drive. The turbocharged engine makes the car a real flyer.*

RIGHT
*The Elan interior has a bright yellow stripe through leather seats. All the controls are to hand and passengers are protected from blustery winds.*

Looking after a new Elan in years to come should be easy. The mechanics are well bred and hopefully full of Eastern reliability. Bodies won't rust, and if the chassis is as well protected as it ought to be in this day and age, the only problem a prospective owner will have will be finding the money. Prices won't dip for a few years until supplies catch up with demand, or a new and more potent version arrives. At present 165 wild horses wouldn't drag true enthusiasts anywhere but to the nearest Lotus dealer.

## PAST CLASSICS

# LOTUS ELAN

The original Lotus Elan began life in the 1960s and won hearts across the world until its demise in 1973. A twin-camshaft 1558cc Lotus engine and independent all round suspension and disc brakes meant the final phase Elans were the best of all. Available as both convertibles and 2+2, the S4 now fetches upwards of £10,000 ($18,500) in tidy trim. The Elan Sprint is another expensive-to-buy car now, having the 126 bhp 'big valve' engine, which would shove the little car to 120 mph (193 kph). Some even carried five-speed gearboxes.

RIGHT
*Later hardtop Elan Sprints are now very pricey, with their 'big valve' 126 bhp engine providing tremendous fun. This car has a popular and evocative two-tone paint scheme.*

# LOTUS ESPRIT TURBO SE

Designed in the early 1970s by the Italian stylist Giugiaro, the Esprit was the first of Lotus's upmarket mid-engined cars, and was never offered as a kit.

By the mid-1970s Lotus had ironed out the teething troubles of faulty cooling, a rickety, vibrating chassis and noisy running gear, and the second series Esprits earned a far better reputation. Starring with Roger Moore in his James Bond escapades naturally helped the image along; the Esprit was the car every successful agent should own, though few were specified with submarine equipment.

In 1983 the first turbos boosted the Esprit into the supercar bracket, and prices took off accordingly. Lotus was bought out (saved?) in 1986 by General Motors, which has generously allowed the small English car maker to pursue an independent development line, which, in this age of motor mogul buy-outs, is indeed laudable.

A budget-priced supercar, clothed in fibreglass, but with a wealth of racing pedigree behind it, the latest and most expensive road-going Lotus, the Esprit Turbo SE, is the fastest car the Norfolk-based company has ever sold. It still doesn't come with the much vaunted active ride suspension, which is sure to be *de rigeur* on all supercars come the later 1990s, but the SE does have a vast amount of power lurking within the 2174cc engine.

Criticized for shoddy build and tacky interiors in the past, the SE could really do with a spot of attention from one of the many talented ergonomists GM must have. Performance-wise you would compare it to a Ferrari or Porsche costing twice as much, and in terms of looks, roadholding and outright desirability, the Esprit is also up there battling away bravely.

LEFT
*The Esprit Turbo SE interior is the height of Lotus sporting luxury. The pale leather cloth is for clean jeans wearers only.*

Designed to be one of the fastest cars Britain has ever produced and engineered using lessons learnt from the oft successful Formula One racing team, the cars are built by a small, dedicated band of craftsmen, keeping up with technology, but not discarding any of the skills of the past.

The current facelift model was introduced at the 1988 Birmingham NEC Motor Show and features better aerodynamics with a more rounded body. The SE has an extra 50 bhp on tap courtesy of General Motors-supplied fuel injection and careful mapping of the turbocharger and engine management system. Suspension is independent all round, with coil springs and multi-links.

Second hand and in good condition they are inexpensive, but not cheap. Invitingly cheap cars should be inspected carefully for cracking and glazing of the bodywork, as well as for crash-damage repairs around front and rear ends, as they're particularly vulnerable here. Chassis have to be straight with so much power available, and if you're buying a supposedly sound car, a dealer service history is a must.

Buying an Esprit Turbo SE makes good sense if you're after a usable supercar, which doesn't cost supercar prices to maintain. Regular maintenance should ensure a long and happy life.

BELOW
*The revamped Esprit debuted in 1988 with more aerodynamic shape and better build quality. Not expensive to buy, the Turbo SE represents great value in the money for performance stakes.*

## TECHNICAL SPECIFICATION

| | |
|---|---|
| **MODEL** Lotus Esprit Turbo SE | **TOTAL PRODUCTION** 2,231 (Esprit models since 1980) |
| **ENGINE** 2174cc, 16-valve, quad cams Garrett T B03 turb, f.i. | **NUMBERS BUILT P/A** c.230 |
| **POWER** 264 bhp @ 6500 rpm/261 lb/ft @ 3900 rpm | **ENGINE/CHASSIS NOS** N/A |
| **CHASSIS** Steel backbone/fibreglass body | **SPARES AVAILABILITY** All spares available |
| **BRAKES** Ventilated discs all round | **PRICE AT LAUNCH** c.£42,000 (c.$77,700); $86,750 (1991) |
| **TOP SPEED** 163 mph (261 kph) | **OPTIONS** Colours: Customer specification<br>Trim: Cloth or leather<br>OE Options: Air conditioning, full leather interior |
| **ACCELERATION** 0–60 mph: 4.7 secs | |
| **PRODUCTION SPAN** 1989– | |

## PAST CLASSICS

# LOTUS EUROPA

A mid-engined two-seater with much racing pedigree, the Europa was powered initially by 1470cc/1565cc Renault engines. Latterly Ford/Lotus twin cam engines found their way under the rear engine deck of the low, fast sportscars, which always used Renault gearboxes. Few were sold as kits, most being built at Hethel, Norfolk, England. Later Twin Cam/Specials had more aggressive styling and trim, alloy wheels and 'Big Valve' engines and are the most desirable. Less than 10,000 Europas were made in all, with desirability ratings being knocked by histories of problematical engines. The Europas are available today at between £5,000 ($9,000) for early S1s and £20,000 ($37,000) for Big Valve S2 Twin Cam 1970s models in good order. American enthusiasts will have to search harder.

ABOVE

*Despite reliability problems with engines, prices for Europas start at around £5,000 ($9,000) for early S1s. Alloy wheels and big valve engines ensure higher prices for the later 1970s S2 Twin Cam models.*

# MORGAN PLUS EIGHT

Morgans haven't seen a serious re-style since before the last World War. Their timeless elegance and traditional sportscar look have kept their popularity high and, with minute quantities produced each year, they are rare and prized cars.

Still built on ash timber frames by experienced craftsmen, each car is constructed to order, with the owner's name on a ticket on each major sub-assembly. Not surprisingly the waiting list stretches five years or more. The Plus Eight has been around almost as long as Rover has been making the GM-derived overhead valve engine; since 1968.

Morgan Cars had developed their four-cylinder engined cars as far as they could and, with the handy alloy V8 on stream offering twice the power with no extra weight penalty, the implementation of the V8 seemed logical.

Widening the track, flaring the wheelarches and smartening up the interior were the only changes necessary. Cast alloy wheels instead of the old wire centrelocks were standard. The old hard-as-a-rock ride continued (to call it a suspension would almost be a crime), but handling was competent enough. Wheelspin in third gear was possible unless the ground under tread was perfectly dry, though with unhelpful aerodynamics, the Plus Eight wouldn't get far beyond 120 mph (193 kph) before the advent of the fuel-injected engine in 1985.

The Plus Eight is easy to service. Engines and gearboxes are still available new. However, the early ash-framed cars had no rot-proofing whatsoever and so are prone to disintegrate. Checking the frame of an older car is therefore essential before buying, and if rot is confirmed, it's expensive to put right.

Since 1986, cars have had their frames soaked in Cuprinol, a proprietary wood preserver, and the undersides are sprayed with an anti-corrosion layer of polyester powder. Some cars were ordered with aluminium bodies so these are the best to buy, since they're lighter, faster and they won't rot out like steel bodied cars will.

The Morgan Plus Eight is collectable because it is a rare, fast and joyful open tourer to travel in. Avoiding bumpy roads soon becomes second nature, but if you see an Eight for sale it won't be cheap. If it is, suspect timber rot and consult an expert. Fuel-injected cars (1985–90) were faster, and 1990 saw the V8 engine bored out to 3.9 litres, as it's now made by Range Rover and fitted to their 4 × 4s. It also has electronic fuel injection.

Some say that hand-built, finely-crafted sportscars are an anachronism in today's world of electro-gadgetary and robot build, but the fact that the cars live on, with such healthy demand, proves that they are a perfectly acceptable form of transport. While the Plus Eight lives on, handfuls of enthusiasts can buy traditional British sportscars for almost a pittance every year.

BELOW LEFT
*With timeless styling, unchanged in almost 40 years and still giving a rock-hard ride, the current Plus Eight cars feature electronic fuel-injected engines in their ash-framed bodies.*

RIGHT
*Crammed under the shapely Morgan snout, the V8 Rover engine could propel the small car so powerfully that wheelspin in third gear was possible.*

## TECHNICAL SPECIFICATION

| | |
|---|---|
| **MODEL** Morgan Plus Eight | **TOTAL PRODUCTION** 3,379+ |
| **ENGINE** 3900cc, V8, OHV, electronic fuel injection | **NUMBERS BUILT P/A** 170 approx. |
| **POWER** c.190 bhp @ 5280 rpm/220 lb/ft @ 4000 rpm | **ENGINE/CHASSIS NOS** 41400002A-47A00026A-R7000-R1039 |
| **CHASSIS** Steel chassis/ash body frame/steel/alloy panels | **SPARES AVAILABILITY** Excellent/ash frame restoration expensive |
| **BRAKES** Discs/drums | **PRICE AT LAUNCH** £1,475 ($2,728); NA |
| **TOP SPEED** 130 mph (208 kph) | **OPTIONS** Colours: Customer specification<br>Trim: Customer specification<br>OE Options: Alloy body, leather upholstery, luggage carrier, walnut dash, coloured hood/mohair hood |
| **ACCELERATION** 0–60 mph: 5.0 secs | |
| **PRODUCTION SPAN** 1969– | |

PAST CLASSICS

## MORGAN 4/4

The Plus Eight grew out of the popular four-cylinder models of the 1960s, which included a four-seater tourer. The 4/4 as it is universally known, is still in production today, still with a Ford 1600cc engine providing up to 88 bhp. Styled like the Plus Eight in another era, the four-cylinder cars have exactly the same loyal following and will continue in production as long as demand holds up. Slightly greater in numbers, though fewer than 3,000 in all were built during the 1970s, the 4/4 is a useful tourer in that it can carry a good amount of luggage and two people, or alternatively four adults in comfort. The 4/4 only loses out in looks when the hood is raised, as the squared-off shape contrasts with the body's swoopy wings and 1930s styling.

RIGHT
*Morgan's 4/4 could carry two extra passengers, though once the hood was raised, the Morgan's lines changed and became less appealing. It is a useful four-seat convertible, though there's a waiting list of up to six years.*

# PANTHER KALLISTA

The Panther Kallista is a direct descendant of the Lima, a Vauxhall-powered open car reminiscent in design of a 1930s sports machine. The Kallista inherited the Lima's looks to a greater extent but under the bonnet Ford engines were moved in. There was a 1600cc unit and a 2.8-litre V6 engine. Later in life the Ford 2.9-litre V6 engine became available, and that is what powered the car up until its demise in the summer of 1990.

The Kallista was killed off by the Panther Solo, not because the latter replaced the £15,000 ($28,000) two-seater, but because Panther needed the production facility and assembly personnel to build their larger, faster, modern supercar.

The Kallista exhibits all the attributes of a traditional English sporting car. It has wide, swooping front wings, a cut-off tail and a pair of large, rounded headlamps, which protrude above the bonnet line and are visible from the driving seat. The cockpit is cosy, albeit too snug for large-framed people, and some of the controls are somewhat tricky to reach. Tall drivers will miss seeing the instruments as the steering wheel rim gets in the way.

Despite all these physical problems, the Kallista is a car to be driven. The rumbling V6 provides great aural entertainment, and the steering is direct and light, effecting snappy directional control. Sitting almost over the rear wheels gives an even more dynamic feel to the effects of steering wheel twirling.

The box section chassis gives excellent passenger protection and the alloy body panels resist rust adequately. There's a snug hood to keep out the rain. The cars were offered with a choice of five-speed manual or four-speed automatic transmission, and were high-geared for relaxed cruising.

What the Kallista offers the classic collector looking for a stylish 1930s look-alike, with sensible, easy-to-source, up-to-the-minute mechanicals, is rarity. It was built by hand, as are all Panthers, and colour and trim specifications were all to the original customer's choosing. Mostly though, there are no garish colour schemes; the Kallista's shape does all the talking.

Sold in 26 countries throughout the world, the Kallista has become well-known and liked. Although the internal accommodation takes some getting used to, the car is basically a useful tourer and fine city dweller.

Replacement parts should not cause any major headaches, since all the running gear is easily sourced from volume manufacturers. The interiors are largely hand-fettled and don't often need attention, and the alloy body panels will be in good supply for many years yet.

ABOVE
*The Kallista's interior is not spacious and the dials are obscured by the steering wheel, but the driving position affords wonderful views along the sweeping bonnet, out over the two large, round headlamps.*

BELOW
*Panther's Kallista was the company's mainstay for six years, taking over from the Lima. Later models with fuel-injected Ford engines are the best to buy.*

## TECHNICAL SPECIFICATION

| | |
|---|---|
| **MODEL** Panther Kallista | **TOTAL PRODUCTION** c.1,800 |
| **ENGINE** 2933cc, V6, electronic fuel injection (Ford) | **NUMBERS BUILT P/A** 250 |
| **POWER** 150 bhp @ 5700 rpm/171 lb/ft @ 3000 rpm | **ENGINE/CHASSIS NOS** N/A |
| **CHASSIS** Steel box section chassis/aluminium body | **SPARES AVAILABILITY** Running gear excellent/body good |
| **BRAKES** Ventilated discs/drums | **PRICE AT LAUNCH** 2.8 litre £7,485 ($13,000) 1983; 2.9 litre £13,875 ($25,600) 1989; 2.9 litre $23,500 (1991) |
| **TOP SPEED** 109 mph (174 kph) | **OPTIONS** Colours: Customer specification Trim: Customer specification. OE Options: Alloy wheels, wooden dash (standard), mohair hood, wood door cappings, wool carpet, metallic paint, duo-tone paint, front spoiler (at extra cost) |
| **ACCELERATION** 0–60 mph: 7.7 secs | |
| **PRODUCTION SPAN** 1984–1990 | |

PAST CLASSICS

## PANTHER DEVILLE

The Panther Deville, built between 1974 and 1985, was a horrible attempt to recreate expensive 1930s styling complete with massive Jaguar engines. Sold at an outrageously high price for the time of around £80,000 ($148,000). only 60 were ever built. The Deville was available as either a saloon or convertible, and ran on straight-six 4.2-litre or V12 5.3-litre Jaguar standard engines.

The cars were opulently finished, though some were done up in garish trim for the rich *sans* taste. The bigger Jaguar engines would pull this overweight monster to 135 mph (216 kph). It is debatable whether these cars are worth big money to the serious investor.

ABOVE
*Outrageously expensive and opulent, the six-door Panther De Ville was an even more horrible attempt at aping times past, but collectable today.*

# PANTHER SOLO

It was in 1987 at the Frankfurt Motor Show that the world first officially clapped eyes on the finished Panther Solo prototype, and a year later the car was launched at the NEC Motor Show in the UK. Finally, in 1990, the first cars were delivered. Months of will-they won't-they worrying for the 100 Britons destined to own one of Britain's most aggressive-looking sportscars, was ended when production ceased after only 14 cars were built by September 1990. With production unlikely to be restarted the 14 Britons who did receive their vehicle should see a very good return on their investment.

The body is a revolutionary composite monocoque of part carbon fibre, part aluminium honeycomb, which is claimed to give the car 'the strongest passenger compartment in the world'. The rear-mounted engine sits in a steel sub-assembly, as does the front transaxle up in the car's nose. Engine power comes from the Ford Cosworth stable, but is only rated at 204 bhp; hardly sufficient, so the testers say, to provide spine-tingling acceleration or top speed. The Solo drives through a five-speed gearbox to all four wheels. The mid-mounted engine proved to be a source of vibration too, which wasn't at all in the Ferrari mould to which the car aspires.

The Solo's dynamic shape does provide something no other supercar can, that's negative downforce at all speeds, meaning that despite the speed at which the car travels, there'll be no lift to distort the steering or blow it off line; the faster it goes, the more it grips the road.

Cornering hard won't produce the kind of over-the-limit, uncatchable spins which afflict most mid-engined cars, and once over the limit, the car will correct its line when the throttle is released.

Even at £40,000 ($74,000) it isn't overly expensive, and does have the looks of a supercar, even if at present the motivating engine power isn't in the supercar league. Panther had the Solo 2 waiting in the wings, and once the first 250 were with their delighted owners, the second phase was to have been financed.

Inside the Panther, leather seats provide good positioning for driver and passenger, and for those who would carry tiny tots in such a car, minute rear seats can be specified. However, they are better left out as they take up what meagre luggage space there is. The front of the car holds no delight for those with sturdy week-end cases, and there are just two handy-sized storage bins in the back, behind the seats.

One thing the Panther Solo is not, is understated. Wherever it goes, mouths will gape. Definitely not for the rock hero seeking anonymity, nor the film starlet out shopping, but, for the celebrity with something to say without uttering a sound, the Solo's the motor.

BELOW
*The Panther Solo took years to reach production, though few will be built each year. Expensive now at around £40,000 (c.$74,000) return on investment depends on whether production is restarted.*

## TECHNICAL SPECIFICATION

| | |
|---|---|
| **MODEL** Panther Solo | **TOTAL PRODUCTION** 14 |
| **ENGINE** 1933cc, 4 cyl, DOHC, electronic fuel injection, turbo | **NUMBERS BUILT P/A** 14 |
| **POWER** 204 bhp @ 6000 rpm/200 lb/ft @ 4500 rpm | **ENGINE/CHASSIS NOS** N/A |
| **CHASSIS** Steel spaceframe/composite body | **SPARES AVAILABILITY** Mech. no prob, body likely to be exp. |
| **BRAKES** Ventilated discs/discs, ABS | **PRICE AT LAUNCH** c.£40,000 ($74,000); not officially exported to North America |
| **TOP SPEED** 150 mph (240 kph) | |
| **ACCELERATION** 0–60 mph: 6.0 secs | **OPTIONS** Colours: Customer specification Trim: Customer specification. OE Options: CD, electric windows and mirrors, central locking, alloy wheels, tinted glass (standard), air conditioning, rear seats, metallic paint |
| **PRODUCTION SPAN** 1990 | |

RIGHT
*The rear view of the Solo is unmistakable. The spoiler detracts from the looks but adds negative downforce at speed, giving greater adhesion.*

PAST CLASSICS

## PANTHER J72

With car production now at a standstill (spring 1991), Panther-seekers will find all models slightly more expensive and exclusive. The J72 was an early model, powered by regular Jaguar six-cylinder engines, though several got the V12 5.3-litre motor. Between 1972 and 1981 over 300 are thought to have been built. The evocative shape aspired to mirror the lavish, luxury sportsters of the 1930s, but fat tyres and overdone, flared wings detracted from the overall impression. Still, they were quite quick, what with the Jaguar engines and light, tubular chassis. Prices as usual are dependent on condition, but for a two-seater with a difference, the J72 is certainly one to think about.

RIGHT
*Suspended production is guaranteed to boost prices of Panther cars, including the two-seater J72, built between 1972 and 1981.*

# RANGE ROVER VOGUE SE

The Range Rover is a trendsetter unique in motoring. The all-wheel drive mud-plugger glamorized off-road, all-terrain vehicles and turned them into fashion accessories for the beautiful people keen for a taste of the country. Exuding virtues of solid construction (steel chassis and alloy bodywork), more-than-sufficient V8-derived power, a high-rise driving position and excellent road manners, the Range Rover became the only vehicle in which to be seen in fashion-crazed Chelsea, South London.

Co-incidentally, the Range Rover is also one of the finest bespoke off-road vehicles money can buy, highly regarded throughout the world, as is its older sibling the Land Rover. Although ignored for much of its early life while the Land Rover division was caught in the machinations of BL Cars, the past ten years have seen an upswing in popularity, linked to its new up-market image.

Accordingly the two-door body style was stretched to four, and though the split rear tailgate was retained, the interiors were adjusted to suit evening wear rather than farmyard boots and jeans. The Range Rover lost none of its good looks converting to four doors, and it is only in the early 1990s that Land Rover are considering major visual revisions.

Designed in the late 1960s to succeed the popular Land Rovers, the Range Rover appealed to the country folk and armies of the world. Only when it moved to be a High Street poseur's delight did the image soften.

The Rover V8 has always been the mainstay motive power, though for different markets, four-cylinder engines, diesels and petrol have been bolted beneath that squared-off sheet of metal in front of the upright windscreen.

Being shaped like the proverbial brick, and having a high centre of gravity led to initial worries over high-speed stability. However the newer models, with their stickier, wider rubber covering smart alloy wheels, handle with the poise of speedy saloons. To police forces in Britain, it proved to be a boon both for off-road work and motorway cruising.

The Vogue trim levels appeared in the early 1980s, plushly fitted out with cloth seats and increasingly voluminous lists of luxury extras. Automatic gearboxes were *de rigeur* and well developed too, more than up to the task, when required, of negotiating ditches and shooting up and over sand dunes.

The SE versions are even smarter with leather upholstery, air conditioning, cruise control and computers. Although the Range Rover is permanently in four-wheel-drive mode, few these days ever venture off the firm grey tarmac. Buying a used example is pretty simple if you

BELOW
*The beautiful people's favourite in-town carriage. To the Vogue SE off-road means parked on the verges of fashionable London suburbs. Prices look set to stay high but the Vogue SE should be good value.*

## TECHNICAL SPECIFICATION

| | | | |
|---|---|---|---|
| **MODEL** | Range Rover Vogue SE | **TOTAL PRODUCTION** | 50,000 |
| **ENGINE** | 3947cc, V8, OHV, electronic fuel injection | **NUMBERS BUILT P/A** | c.10,000 |
| **POWER** | 185 bhp @ 4750 rpm/235 lb/ft @ 2600 rpm | **ENGINE/CHASSIS NOS** | N/A |
| **CHASSIS** | Steel/some aluminium body panels | **SPARES AVAILABILITY** | All parts available |
| **BRAKES** | Ventilated discs/discs/specially developed ABS | **PRICE AT LAUNCH** | £20,000 ($37,000); N/A |
| **TOP SPEED** | 112 mph (179 kph) | **OPTIONS** | Colours: Ardennes green, arles blue, clearwater blue, plymouth blue, westminster grey<br>Trim: Leather upholstery. OE Options: Heated front windscreen, central locking, (Vogue SE) WABCO-designed ABS |
| **ACCELERATION** | 0–60 mph: 9.9 secs | | |
| **PRODUCTION SPAN** | 1983– | | |

BELOW
*Inside the Vogue SE there's leather trim and lots of electric goodies to play with. Smart carpets too are a Vogue specification feature.*

visit a main dealer in an urban area. Models sporting towbars need not necessarily indicate hard lives on farms, though horsebox lugging is definitely a possibility.

Even after 20 years constant production, and ten years continuous upgrading, the Range Rover is instantly identifiable, and with aluminium body panels, generally bodily sound. Checking chassis for straightness and engines for stump-pulling power are the only real investigations you have to make. Still ahead of the Japanese cheaper imitations, the Range Rover will be revered long after they have rusted away.

## SPECIAL EDITIONS

### RANGE ROVER CSK

To celebrate 20 years of the Range Rover, makers Land Rover issued a rare limited edition of 200 two-door CSK anniversary models in the autumn of 1990. Named after the off-roader's designer Spen King, and identifiable by their Beluga black paintwork, chromed bumpers, extra driving lamps and a special plaque confirming the model's exclusivity, the four-wheel drive vehicles could top 114 mph (183.5 kph).

The CSK used the latest 3.9-litre V8 185 bhp engine, fed by electronic fuel injection and stopped with the aid of four-channel anti-lock brakes in extremis. Inside, the CSK was trimmed to the highest Range Rover standards, with beige leather seats, door trims and steering wheel. American walnut inserts in the door cappings and fascia, and a tilt/slide sunroof are other CSK refinements. Externally the special model used four distinctive, black-edged, silver five-spoke alloy wheels to round off a unique, and in years to come collectable, Range Rover priced at launch at £28,995 ($53,000) as a five-speed manual, or £30,319 ($56,000) with the fuss-free automatic gearbox.

RIGHT
*With a top speed of 114 mph (184 kph), the CSK is the fastest ever production Range Rover. At £28,995 ($53,000) it is also the most expensive.*

# ROLLS ROYCE CORNICHE III CONVERTIBLE

It's been in production longer than any other current Rolls Royce, and to some it is their finest offering. The property developer's dream machine in the early 1970s, the hand-built drop-head Roller was the status symbol every well-heeled *nouveau riche* gent desired. Now though, their purchase price heads easily into six figures, and the value of older cars shoots upwards. Don't be surprised if poorly maintained cars appear for sale well over their original purchase price.

When the Silver Shadow lost its roof and two rear doors, the resulting shape was a marvellous rendition of a popular theme. Through the 1950s and 1960s convertible Rolls Royces were the desire of society's *creme de la creme,* and the car to be seen in driving through Knightsbridge, down to Monaco, or the French Riviera for the weekend.

Impeccably turned out to the last, the Corniche convertible has always been a very rare car, made almost to order. The cars are still built today at Mulliner Park Ward coachbuilders in North London.

That it survives bears testimony to its looks and character. No one doubts that Rolls Royce could have done a similarly effective job with the Silver Spirit/Spur, but few could argue that it would have been with such style.

Luckily, impending US regulations at the time of the Spirit/Spur design directed the drawing office away from the convertible option and the Corniche lived on. The car was available as a two-door hardtop too, a popular option, also sharing the familiar 6.75-litre V8 engine.

As solid without a roof as with, the Corniche travels like a luxury tourer, unruffled when unhustled, but composure at speed needs concentration. In USA guise the ride is less than firm, and rather too wallowy for many. However, as

LEFT
*Inside the Corniche, comfort reigns supreme; it's the biggest British convertible you can get.*

BELOW
*Elegance and sumptuous comfort identify this long-running drop-head. Find a pure, untouched early model or buy a brand new one!*

## TECHNICAL SPECIFICATION

| | |
|---|---|
| **MODEL** Rolls Royce Corniche III Convertible | **TOTAL PRODUCTION** N/A |
| **ENGINE** 6750cc, V8, digital fuel injection | **NUMBERS BUILT P/A** c.300 |
| **POWER** 233 bhp @ 4300 rpm/332 lb/ft @ 1600 rpm (unoff. figs) | **ENGINE/CHASSIS NOS** N/A |
| **CHASSIS** Unitary steel, separate front, rear subframes | **SPARES AVAILABILITY** Excellent/expensive |
| **BRAKES** Ventilated discs/discs, ABS | **PRICE AT LAUNCH** £13,439 ($25,000); $220,100 (1990) |
| **TOP SPEED** 130 mph (208 kph) | **OPTIONS** Colours: Customer specification<br>Trim: Customer specification<br>OE Options: Air conditioning (standard) |
| **ACCELERATION** 0–60 mph: 9.0 secs | |
| **PRODUCTION SPAN** 1971 to present | |

a boulevard cruiser, the car is possibly without compare, and it is this fact that has kept up its popularity across the water in the USA.

Fitted with all the latest electronics and improved componentry as it was applied to later Rolls Royce cars, the 1990s version is as up to date as is possible given that the original design is well into its twenties. Automatic transmission is standard, as is air conditioning.

A new Rolls Royce convertible is rumoured, and when the Corniche dies, an era will go with it. Cars which have been well cared for will obviously command premium prices. But for enthusiasts willing to get down to some hard work, there are cars around at below average prices. The luck of the draw is finding a neglected example – more likely in Europe, Florida or California than in the UK – and then restoring it slowly to former glory.

A brilliant car in its day, and still in demand after 20 years, the Corniche is a true classic, and highly collectable.

### PAST CLASSICS

## ROLLS ROYCE SILVER CLOUD CONVERTIBLE III

Any Rolls Royce convertible from the 1960s is bound to be a great car, both as a running classic and as an investment. Cars based on the Silver Cloud chassis were coachbuilt by several noted body designers, including H J Mulliner, Hooper, Freestone & Webb and James Young. The Mulliner/Park Ward-designed drop-head (type 2045) is the most popular and the most elegant. Complete with twin slanting headlamps (a controversial choice for Rolls Royce), the car was sold both as a Rolls Royce and a Bentley and was built in very small numbers; only 84 Bentley and even fewer Rolls Royce convertibles were crafted between 1963 and 1965. All cars ran the 6250cc V8 engines and cost £7,000 at launch. Where the cars differed was in the radiator shells and gauge layout for ancillaries on the dashboard. Pristine condition cars are valued today at around £100,000 ($185,000).

RIGHT
*The Silver Cloud Convertible was sold both as a Rolls Royce and as a Bentley, however production was still minute.*

# TRIUMPH TR8 CONVERTIBLE

The Triumph TR8 is the car that never was, at least not for the UK market; the Americans had more luck, having had the car on sale for two years. Back in the loss-making, industrial-heartbreak days of the early 1980s BL Cars were in a tight spot. Having moved TR7 production literally all over the country, the V8-engine cars seemed on the brink of mass production when the entire TR7/8 range was axed. The pretty convertibles and close-coupled coupes with Rover V8 engines tightly packed under broad, flat bonnets never saw the volume production they rightly deserved.

The TR8 was simply a planned production evolution model of the TR7 range, a radically styled but impossibly badly built package. The doors didn't shut, the pop-up headlamps didn't and mechanical infirmity gave the cars a bad name.

Everyone had expected more power when the V8s were born and with 155 bhp quoted for the UK market, they weren't to be disappointed. In the US, emission equipment reduced power output to 135 bhp. Cars were sold with carburettors or fuel injection. Suspension was stiffer than on the TR7s, power steering was standard, and the battery sat in the boot.

BELOW

*Triumph's TR8 was the last of a famous line of British sportscars to take America by storm. A handful were sold in Britain, but fakes abound. The underbonnet hid Rover's alloy V8, which could pump 155 bhp, if not smothered by emission control equipment.*

The beauty of the TR8 was that it needed no option list of any importance. You didn't need a turbo with the prodigious torque of the V8, which hit peak power at a lowly 2,500 rpm. Acceleration was brisk enough for the most sporting of tastes and for round-town work 2,000 rpm was more than adequate.

The TR8 would have found many friends among die-hard TR people, who by 1981, had got over the wedge-shaped styling shocker of 1975. Apart from rarity in the UK, the TR8 is an excellent car to buy as a classic, since it offers superb performance from an unfussy powerhouse. Spares will always be plentiful since it shares many components with the TR7 and the running gear is still in production in some shape or form.

The car was never officially offered in the UK; the only cars which are around are from the handful of press demonstration cars which were subsequently sold off after the project died.

Prices will always be high for the handful available, but beware of cheap imitations. Many TR7s have been converted to V8 power and are not pukka TR8s. These interlopers are of little except curiosity value, and a real V8, especially the convertible, is one to be prized.

BELOW RIGHT

*The TR6 was a squarer development of the TR4/5 cars, complete with the 2500cc six-cylinder engine. Smart cars now fetch £10,000 (c. $18,500).*

## TECHNICAL SPECIFICATION

| | |
|---|---|
| **MODEL** Triumph TR8 Convertible | **TOTAL PRODUCTION** 2,722 (most to USA) |
| **ENGINE** 3528cc, V8, OHV | **NUMBERS BUILT P/A** c.1,500 |
| **POWER** 155 bhp @ 5250 rpm/198 lb/ft @ 2500 rpm | **ENGINE/CHASSIS NOS** N/A |
| **CHASSIS** Unitary steel | **SPARES AVAILABILITY** Good to excellent |
| **BRAKES** Discs/drums | **PRICE AT LAUNCH** c.£12,000 ($22,200); NA |
| **TOP SPEED** 140 mph (224 kph) | **OPTIONS** Colours: Standard BL cars<br>Trim: Cloth checkered seat coverings<br>OE Options: Air conditioning |
| **ACCELERATION** 0–60 mph: 8.2 secs | |
| **PRODUCTION SPAN** 1980–1981 | |

PAST CLASSICS

## TRIUMPH TR6

No TR previous to the TR7/8 had ever been so dramatically styled or modern-looking. In fact the TRs had stood for traditional conservative English sporting heritage. The car the TR7 replaced was the conventional and quite pretty looking TR6, powered by the Triumph 2.5-litre engine, and available with petrol injection as an option.

The TR6 suffered the appalling build problems which were later to afflict TR7/8s. The low-slung, open two-seater body was aggressively styled and tuning experts claimed to be able to extract 150 bhp from the 2.5-litre six-cylinder lump. For some reason prices for good TR6s have reached £10,000 ($18,500) in the UK, though it's difficult to see why, since almost 100,000 were built between 1969 and 1976.

# TVR 450 SEAC

Built by hand in the seaside resort of Blackpool, TVRs have for several years, been making very good use of Rover's alloy V8 engine. Bored out to several different capacities and always highly tuned, the latest 450 SEAC takes the concept to its limit, complete with Cosworth engine modifications.

SEAC stands for Special Equipment Aramid Composite, which means that the body is a Kevlar composite, giving light weight but excellent strength and crash-damage resisting features.

Available as a coupe or convertible, the tubular steel backbone chassis takes the strain of almost 320 horses charging around wildly under the fibreglass engine cover. Occupants sit almost over the rear axle which takes the drive, and the long, disappearing nose makes judging distances difficult. Minor inconveniences for a car which offers stunning open-road performance, but requires regular replenishment at the pumps.

Rarity is this car's greatest asset, and don't be confused by the plethora of like-coded lesser models; there were the 350, 390, 400 and 420 models before the 450, so watch the badging. Options on these cars were to a high standard, so you might find leather seats, quality radio systems, and electronic widgets everywhere. If you're looking to buy, ensure that all the toys function as they should, since they'll be expensive to replace, unless of course they're the bits from the parts bin of a large manufacturer.

Unfortunately, build problems in later cars meant that some panels dropped off occasionally, front spoiler-bumper sub-assemblies in particular. This often resulted in the low-slung exhaust being ground over uneven terrain.

TVR have only built 38 examples thus far, but of the other derivatives with basically similar body structure, there are many more. All these TVRs offer excellent performance and ride characteristics; great British sporting cars complete with traditional British hand finishes. Take off the rose-tinted specs though, when viewing a potential purchase.

BELOW
*TVR's biggest-engined car had a small market, but the fibreglass-bodied V8, with its track breeding, had plenty of pace. These cars are expensive but rare, fast and rewarding, even if somewhat noisy.*

## TECHNICAL SPECIFICATION

| | |
|---|---|
| **MODEL** TVR 450 SEAC | **TOTAL PRODUCTION** 38 |
| **ENGINE** 4441cc, V8, electronic fuel injection | **NUMBERS BUILT P/A** approx 9 |
| **POWER** 319 bhp @ 5700 rpm/310 lb/ft @ 4000 rpm | **ENGINE/CHASSIS NOS** N/A |
| **CHASSIS** Tubular steel backbone/fibreglass body | **SPARES AVAILABILITY** All parts available |
| **BRAKES** Ventilated discs all round | **PRICE AT LAUNCH** £28,000 ($51,800); NA |
| **TOP SPEED** 160 mph (256 kph) | **OPTIONS** Colours: Customer specification<br>Trim: Customer specification<br>OE Options: Leather interiors, electric windows and mirrors, quality radio cassette |
| **ACCELERATION** 0–60 mph: 4.7 secs | |
| **PRODUCTION SPAN** 1986– | |

RIGHT
*Interiors are to personal choice, but include Wilton carpets, half or full leather trim, wooden dash inserts and leather steering wheel. The view down the long, sloping bonnet is hugely impressive. Electronic features include windows, mirrors and central locking. The cosy two-seater cabin is well heated, but listening to the V8 growl must be the biggest interior thrill.*

PAST CLASSICS

## TVR TUSCAN V8 SE

The TVR Tuscan V8 SE, launched in 1968, could be a forebear of the current top-of-the-range SEAC. Only 21 were ever built, but this Mustang-engined rocket would top 140 mph (225 kph) with 271 bhp on tap. It was a fearsome brute, although the svelte little body, all bonnet and stubby tail, belied the power under the hood. Centre-lock wire wheels and slightly longer wheelbase are the only external give-aways to a car which closely resembled the smaller-engined Tuscan V6, which was powered by a Ford V6 from a Capri, and was capable of 125 mph (200 kph).

RIGHT
*TVR Tuscan used Ford's Capri V6 3-litre engine managing 140 mph (225 kph), so it's no snail. Values today aren't out of reach, but Tuscans are hard to find.*

# HONDA NSX

After many years producing some of the most attractive of Japanese bread-and-butter, up-market saloons, Honda are finally putting the lessons learnt in Formula One to good use in an excitingly styled luxury car. The likes of Porsche and Ferrari should sit up and take note. The NSX is specifically designed to be the world's most sophisticated sports machine, but one which can be driven by just about anyone.

Aluminium is a key feature of the new car, making up the body, engine and suspension componentry. Inside there's the usual range of Japanese goodies to keep the drivers amused. But under the bonnet is where the action really is.

The 2977cc V6 pulls the NSX all the way to 165 mph (265 kph), courtesy of the variable valve timing system, unique to Honda's engines. At low speeds cam lobe settings act to give slightly out-of-phase openings on one of the two inlet valves. This produces more swirl and thus better combustion inside the cylinders. At higher revs the lobe profile changes and the valves open simultaneously, as the camshaft takes on a harder profile. While not yet available in the UK complete with turbo, the NSX is a fine performer running on fuel injection alone.

The NSX will score many points for its styling, being one of the prettiest serious sports machines the Japanese have so far produced. Even at around £50,000 ($92,500) the car will find many admirers, and, although production will be more than a mere trickle, the numbers entering the UK will not be vast; American buyers will be far luckier.

Designed to take on the Porsche 911 and the Ferrari 348, the Honda will find it a hard task unseating two makers who have for so long had the reputation for marvellous high-performance cars. Honda don't yet have that certain cachet which turns drooling enthusiasts into wallet-opening customers.

But for the wealthy individual who wants to be different, while keeping up with the speed merchants, the NSX will have the novelty of rarity for the next couple of years at least. The Japanese are at last conceptualising niche marketing, and as long as they don't flood the market with thousand upon thousand NSXs, price, rarity and image will make the NSX the dream car Honda have been looking for.

RIGHT
*The Honda NSX captures the essence of supercar travel but in a remarkably docile manner. Expensive and in short supply in the UK, the US will get 3,000 plus cars every year.*

ABOVE
*The mid-mounted transverse engine has variable valve timing and a host of advanced mechanical goodies, but with traditional Honda reliability. The luxury of the interior is intended to match European rivals.*

Rarity will keep prices buoyant for a long while to come, and with Honda's spares system, no car should ever be stranded for want of a part. Looks enough will guarantee a healthy interest in owning one, and until the Japanese manufacturers *en masse* bring their concept cars into full-scale production, the NSX will be the Far Eastern performance package to go for.

Even when the likes of Mitsubishi have their sports coupes rolling from the lines, as they most definitely will in due course, the Honda NSX will go down as a brave attempt to get the combination right. They have so nearly achieved their goal; their next generation will certainly be worth waiting for.

The NSX is sold in the USA as an Acura, Honda's stateside brandname.

## TECHNICAL SPECIFICATION

| | |
|---|---|
| **MODEL** Honda NSX | **TOTAL PRODUCTION** N/A |
| **ENGINE** 2977cc, V6, 24 valves, DOHC, V-Tech | **NUMBERS BUILT P/A** 4,500 (3,000 to USA) |
| **POWER** 274 bhp @ 7100 rpm/209 lb/ft @ 5400 rpm | **ENGINE/CHASSIS NOS** N/A |
| **CHASSIS** All aluminium monocoque | **SPARES AVAILABILITY** All parts will be available |
| **BRAKES** Ventilated discs all round | **PRICE AT LAUNCH** c.£52,000+ ($96,000); $60,000 (5-speed manual) |
| **TOP SPEED** 165 mph (264 kph) | **OPTIONS** Colours: Red, silver black<br>Trim: Leather. OE Options: Automatic transmission, air conditioning, traction control system, air bag, central locking, cruise control; co optional |
| **ACCELERATION** 0–60 mph: 6.0 secs | |
| **PRODUCTION SPAN** 1990– | |

PAST CLASSICS

## HONDA S800

Potentially the most exciting car Honda had ever made up until the 1980s, the 1966-introduced S800 sportscar won the hearts of car enthusiasts the world over. Styled in the manner of the MG Midget, the S800 ran four carburettors on its 791cc engine which was a four-cylinder unit complete with double overhead camshafts. The somewhat revy engine produced about 70 bhp and the four-speed gearbox was delightful. The S800 was available, like the MGB, in roadster or closed coupe form, and is now considered a collectable classic.

RIGHT
*Honda competed with MG with two pretty, little sportscars, both fitted with tiny jewel-like 800cc engines and fed by four carburettors. Coupe and fastback styles are sold, and the cars are now highly collectable.*

# MAZDA MX-5

The Mazda MX-5 is the most raved-about two-seater sportscar for decades. Not that it is a performance hotshot, nor is it fabulously expensive. It is the modern-day Lotus Elan, the car which carries through the spirit of freedom, sporty character and handling, while being sensibly priced and easy to drive.

Comparisons with the early Lotus Elans have been made ever since the first prototypes poked their plastic noses into the world. The MX-5 has been marketed as a car to take you back to the good old days of the 1960s, jumping quite unashamedly on the nostalgia bandwagon. However, like all things Japanese, the MX-5 has been fully developed before it reached production, not during its life, thus ensuring reliability, a key concept in sportscar terms.

The MX-5 is powered by Mazda's 323 engine, which has been reworked slightly to give a few extra revs at the top end, and fitted with electronic ignition and fuel injection. Tuned to a modest 114 bhp, the MX-5 won't strip the tread from the tyres leaping into the distance, but it will handle like a zippy two-seater should, quite neutrally and utterly controllably.

In the USA the cars were so much in demand after launch that the initial selling price was more than doubled by black market investors. They tried to off-load early cars to desperate and monied prospective MX-5 owners.

Thankfully the supply of cars has increased and £27,000 ($50,000) no longer changes hands to secure one. In the UK, Mazda is currently thwarted by an import quota and can only offer the bewitched public 2,500 units per year.

Practical to the last, the hood can be quickly raised in the event of the all-too-common showers, without having to leave the driving seat. What's more, the operation only takes about 10–15 seconds with practice. The plastic nose section is strengthened to absorb bumps.

Pop-up headlamps and a modest bonnet bulge are the only interruptions to a bland front end, but from the side, the MX-5 looks impressive. Details such as the finger latches copied from the Alfa Duetto, and Elan-style rear circular tail lamps show Mazda studied its market carefully.

The MX-5 will continue to be popular while the public perceives it to be exciting and novel. Once the initial euphoria dies down, there will be stronger cries for more power. Hopefully that won't spoil the inherent character of a perfectly balanced sportscar.

The MX-5 is a classic because it faithfully copies the traits of previously successful, small sportscars. It's nippy, handles well and being so solidly engineered, it shouldn't develop any sort of reliability problem. The MX-5 is cheap enough to appeal to a broad market and prices will hold firm for several years to come.

BELOW

*Mazda captured the hearts of all small sportscar lovers with their late 1980s Miata/MX-5. Styled unashamedly on the old Lotus Elan, the car provides sporty fun at a reasonable price.*

## TECHNICAL SPECIFICATION

| | |
|---|---|
| **MODEL** Mazda MX-5 (MIATA) | **TOTAL PRODUCTION** 66,710 to date |
| **ENGINE** 1598cc, 16 valves, DOHC, fuel injection | **NUMBERS BUILT P/A** 2,500 sold in UK 1990 |
| **POWER** 114 bhp @ 6500 rpm/100 lb/ft @ 5500 rpm | **ENGINE/CHASSIS NOS** N/A |
| **CHASSIS** Unitary steel/alum bonnet/plastic nose and tail cone | **SPARES AVAILABILITY** All parts available |
| **BRAKES** Ventilated discs/discs | **PRICE AT LAUNCH** £14,249; $13,200 (1991) |
| **TOP SPEED** 114 mph (182 kph) | |
| **ACCELERATION** 0–60 mph: 9.1 secs | **OPTIONS** Colours: Standard Mazda shades. Trim: Cloth seats OE Options: Alloy wheels, electric windows, removable cassette player |
| **PRODUCTION SPAN** 1989– | |

BELOW
*Under the hood the MX5 hides a slightly tuned 1600cc engine. New more powerful engines are on the way, but the current unit does a nice enough job.*

## SPECIAL EDITIONS

## MAZDA MX-5 LIMITED EDITION

Being so overtly based on the Lotus Elan, the MX-5 has no classic Mazda antecedents. However, ever keen to capitalize on Japanese infatuation with small British sports cars, Mazda launched the MX-5 Limited Edition in the summer of 1990. Not scheduled for the UK until 1991, the Limited Edition comes in British Racing Green, has a wood-rimmed steering wheel and wooden gear knob, as well as alloy wheels, electric windows and CD player. Tan leather seats complete the interior package. It's as close to a Lotus Elan copy as Mazda can go. Mazda aim to sell 500 of the Mazda MX-5 Limited Edition each month in Japan at a price of around £8,500 ($15,750).

JAPAN

# MAZDA RX7

If you were to call the Mazda RX7 a true 1980s sportscar, it might seem like something of a slight. It shouldn't, since Mazda were bold enough to build and successfully sell an overtly sporting hatch at a time when few others were interested. But more significantly, they were the only people in the world with sufficient confidence to power it with a rotary engine.

The earlier cars were sharply styled, often garishly coloured but undeniably effective performance machines. The hatch was useful and did not detract from the sporting appeal of the shape. In 1986 the concept was overhauled with the series two car, which has since sprouted a very deftly styled convertible, as well as a turbocharged version of the rotary engine.

The Mazda RX7 is a high-performance tourer in either body style, ably gobbling up miles on point-to-point runs. Although the rotary engine is not known for fuel economy, the performance from the double rotor engine is phenomenal. The turbocharger was specially designed for the RX7 to minimize lag and increase turbine efficiency. The exhaust ports are split and differently sized to allow better control of entry gases into the turbo.

The RX7 attains classic status largely because of the unusual engine design, and also because it is a fine-looking sports car. The sleek front end slices the air cleanly and the windblocker on the convertible neatly protects passengers from the effects of wind turbulence at high speed.

The RX7 is a finely balanced sportscar, which is aimed at the high-performance tourer category, practically in the Porsche 944 league. When you study the profile of the coupe you'll see in which direction the Mazda engineers were looking when they drew up this Japanese offering.

Where the convertible scores is by having an almost rattle-and-shake free tub, and an appearance to match any car at twice its price. The rag-top is well proportioned and looks fast and sporty even at rest. The powered hood works nicely with its targa centre section, but raising or lowering the roofing takes minutes and many different phases to complete.

The RX7 has an inviting cockpit, and once seated behind the wheel, the large dials stare helpfully back. The rake of the broad windscreen is quite steep. Leather seats enhance the luxury and quality-of-build feeling, and the short stubby gear lever falls perfectly to hand.

ABOVE *The Mazda RX7 is available as a coupe or convertible and offers sling-shot performance from its rotary engine. Earlier cars are good value if the engine is in good condition.*

The RX7 will enter the history books as a well-developed sports-hatch, competently powered, elegantly designed and rare enough to earn a second glance. Spine tingling acceleration from the turbocharged rotary leaves an indelible impression. Every true sporting enthusiast should have one. They're not expensive to buy second hand, and traditional rotary phobia can be ignored on dealer-serviced, well-maintained cars.

## TECHNICAL SPECIFICATION

| | | | |
|---|---|---|---|
| **MODEL** | Mazda RX7 Coupe/Convertible | **TOTAL PRODUCTION** | c.400,000 |
| **ENGINE** | Twin rotary 1308cc, turbo, fuel injection | **NUMBERS BUILT P/A** | 39,450 (1990) |
| **POWER** | 182 bhp @ 6500 rpm/183 lb/ft @ 3500 rpm | **ENGINE/CHASSIS NOS** | N/A |
| **CHASSIS** | Unitary steel | **SPARES AVAILABILITY** | All parts available |
| **BRAKES** | Discs/discs | **PRICE AT LAUNCH** | £22,000/£24,000 ($40,000/$44,000); $26,700 (convertible 1991) |
| **TOP SPEED** | 135 mph (216 kph) | | |
| **ACCELERATION** | 0–60 mph: 7.2 secs | **OPTIONS** | Colours: Standard Mazda shades<br>Trim: Leather<br>OE Options: Air conditioning |
| **PRODUCTION SPAN** | 1985– (convertible 1989–) | | |

ABOVE
*The RX7 convertible has a unique 'windjammer' to prevent wind ruffling the passengers' hair. Even if it is expensive, it's a good bet for the future, especially with that rotary motor.*

PAST CLASSICS

## MAZDA RX 3

Arguably the most famous and popular, sporty Mazda is the RX 3, a two-door coupe, twin-rotary, wankel-engined design, produced between 1971 and 1978. Over 280,000 were built, attesting to the car's desirability. Complete with Americanized frontal styling, the car looked effective, though not as brutal as real American cars of the period. Up to 120 bhp was on offer, though top speed was limited to just over 100 mph (160 kph). The RX 3 is underrated at present with rust a major killer but, for future reference, may prove a good buy.

RIGHT
*The two-door version of the popular RX 3 found few friends in the UK. Today most have rusted, but they make interesting classics, and they are quick.*

# NISSAN 300ZX

If there were ever a jewel in the Nissan crown during the 1970s it was the 240Z or, as it was known in the USA, the Fairlady. Nissan were guilty of producing some really unimpressive cars at that time, but took the world by storm with the 240Z. It was fast, good-looking, handled well and was cheap. In that all-important American market it sold for around $3,500 (£1,900), offered everything and more than anything the British could serve up, and it suited the market perfectly.

Nissan built over 600,000 240Z and 260Z examples. Like all successful cars, the Zs suffered from over-development in the following years which culminated in the rather uninspiring 300ZX, with its sharp nose and spongy US-spec suspension.

In 1989 a new 300ZX appeared, shaped for the 1990s and modelled on the smaller Porsches and Ferraris, as well as carrying over the spirit of the 240Z. Aimed at the lucrative, up-market coupe sector, the 300ZX offers plenty of driver interest with a 3-litre V6 engine, complete with obligatory 24 valves, four camshafts, twin turbos and four-wheel steering. The new 300ZX is, however, far more expensive, costing around £35,000 ($65,000) in the UK.

With all this up-to-the-minute gadgetry it will be easy to haul the punters into the showrooms, and once they see the shape, they'll be hooked. The 300ZX has *presence,* something few Japanese cars have ever had. From the sleek, aerodynamic nose, over the steeply raked windscreen to the stubby, rounded tail the shape looks modern and exciting.

The 300ZX will carry the torch of desirability, relinquished in the late 1970s by the last real Z sportscar. Since that time, ever-increasing emissions controls have deprived the ZXs of true sporting ability.

Smaller-than-average twin turbos prevent unnecessary lag and additionally, variable inlet valve timing increases response at the bottom end.

The turbo and T-bar versions will be the most popular options in markets where they're

BELOW LEFT

*A purer, more sporty 300ZX from Nissan looks and performs far better than the last square 300, and it will be a car to savour.*

offered, and the turbo will provide outstanding performance; the sequential fuel injection on the standard car is sufficiently satisfying as it stands at the moment.

Priced so far into Porsche territory, the 300ZX supercar may never be the volume seller of its predecessor, but that won't prevent it being a force to be reckoned with. Those lucky enough to own a car of this ilk will appreciate the 300ZX if they drive it hard enough, and subsequent owners will find themselves with a performance package undaunted by a few years of use. Nissan trades on a reliability record most other manufacturers dream about.

## TECHNICAL SPECIFICATION

| | |
|---|---|
| **MODEL** Nissan 300ZX | **TOTAL PRODUCTION** N/A |
| **ENGINE** 2960cc, V6, 24 valves, elec. fuel injection, twin turbos | **NUMBERS BUILT P/A** N/A |
| **POWER** 280 bhp @ 6400 rpm/274 lb/ft @ 3600 rpm | **ENGINE/CHASSIS NOS** N/A |
| **CHASSIS** Unitary steel | **SPARES AVAILABILITY** All parts available |
| **BRAKES** Ventilated discs all round, ABS | **PRICE AT LAUNCH** c.£37,500 ($68,000); $27,000 |
| **TOP SPEED** 153 mph (245 kph) | **OPTIONS** Colours: Standard Nissan range<br>Trim: Cloth, electric leather seats optional<br>OE Options: Standard features automatic gearbox, metallic paint, extras: ABS, alloy wheels, power steering, air conditioning cruise control, electric windows, central locking |
| **ACCELERATION** 0–60 mph: 5.6 secs | |
| **PRODUCTION SPAN** 1989– | |

## PAST CLASSICS

# DATSUN 240Z

The 240Z was launched in 1969, and took over a market so long dominated by British sporting cars the British viewed the category as their own. The 240Z was fast, reliable and practical. It handled well and drove like a charging sportscar should. Unhelpfully, subsequent versions became softer in the wake of American demands, but even as late as 1983, when the 280ZX was phased out, there were discernible sporting pretentions left. The 300ZX, which soldiered on to 1989, was a pale shadow of the 240Z in terms of performance, though it was still good value.

RIGHT
*The original Z car, the 240Z outsold the Jaguar E-type and became immensely popular in the US. Later versions however became ponderous and heavy, and were handicapped by emission controls.*

JAPAN

# TOYOTA MR2

This is Toyota's second crack at mid-engined, budget sportscar. The first-generation MR2's wedge-shaped body won it many friends, even if the 1558cc engine was a little lacking for ultimate speed. As with most developments, the new car is appreciably bigger (wheelbase has grown by just over three inches, and overall length by 8½ inches). Despite this middle-age spread, the new MR2 has a choice of 2.0-litre engines to spirit it along, and for the US market there's a 2.2-litre motor to cope with the strangulation enforced by emission controls.

Mid-engined cars always lack significant luggage space, but the new MR2 has room for a set of golf clubs and their hold-all.

Targeted at the young, sporting market, the car exhibits lovely ride and handling characteristics. There's nothing to worry about in the normal course of events and, with 158 bhp on tap (in UK guise) or 119 bhp for the normally aspirated version, the MR2 is pretty adventurous. The turbo engine is basically the same unit as used in the top-of-the-line Celicas and features 16 valves and twin overhead camshafts.

Pop-up headlamps and beautifully rounded curves give a very Ferrari-like impression, especially when viewed from the side. The versions carrying the rear spoiler don't have their overall looks too compromised in the search for extra rear downforce.

Rear-brake cooling slots, forward of the rear wheelarches, add an exotic air to the overall appearance, and handsome alloy wheels are shod with low profile tyres. The combination of these pocket-exotic, less angular looks and Toyota engineering result in a car to be sought after and enjoyed by anyone with high driving expectations.

The driving position is excellent, the curved dash giving the driver easy access to all the controls, and the cabin is cosy as well as sporty. Despite the acres of plastic which is a traditional Japanese feature, the overall impression is of quality not crass execution as in so many previous offerings from the Far East.

The new MR2 has managed to retain its cheeky looks and characterful performance and will be a sought-after car. The later turbo versions undoubtedly will be most attractive to the classic collector.

BELOW LEFT
*Toyota improved the square MR2 in 1990 with a softer and more powerful successor. Even in non-turbo form the car handles and performs well for a mid-engine car, and will be one to look out for.*

LEFT
*Inside, the new MR2 caters well for drivers; all the controls are to hand, and the cabin is snug and welcoming.*

The shape of the new MR2 makes it an even more practical sportscar, but the earlier car was a trendsetter for its age. Where the new MR2 will catch the classic enthusiast's eye is in the styling department. Toyota cleverly seized upon customer enthusiasm and re-developed their mid-engine budget-sized layout to meet demands for a bigger, more airy cabin, whilst revamping the dated mid-1980s external panelwork. The result is an affordable and aerodynamic supercar.

## TECHNICAL SPECIFICATION

| | |
|---|---|
| **MODEL** Toyota MR2 | **TOTAL PRODUCTION** 166,000 (old and new shape) |
| **ENGINE** 1988cc, mid-mounted, electronic fuel injection | **NUMBERS BUILT P/A** 45,425 (old and new shape) |
| **POWER** 158 bhp @ 6000 rpm/140 lb/ft @ 4800 rpm | **ENGINE/CHASSIS NOS** N/A |
| **CHASSIS** Unitary steel | **SPARES AVAILABILITY** All parts available |
| **BRAKES** Ventilated discs all round | **PRICE AT LAUNCH** c.£15,441; estimated $24,000 |
| **TOP SPEED** 137 mph (214 kph) | **OPTIONS** 119 bhp models available in UK only, with electronically controlled automatic four-speed and manual five-speed gearboxes. This downspec model uses the Toyota Camry 2.0-litre engine and cost £14,000 at launch. US spec 2.2-litre models have 16 valves, twin camshafts and fuel injection. Turbo models are on the way for 1991. Colours: Standard Toyota shades. Trim: Cloth seats OE Options: Leather seats and T.-bar roof extras |
| **ACCELERATION** 0–60 mph: 6.7 secs | |
| **PRODUCTION SPAN** 1990– | |

## PAST CLASSICS

## TOYOTA MR2

The old MR2 was offered with a 1587cc engine and had far less luggage space than its successor. The T-bar roof option was very popular, the majority of the cars sold in the UK and US being so equipped. Launched in 1984, it competed with the Bertone X1/9 and shared a large market of mid-engined enthusiasts. The MR2 used its 1587cc capacity to the full, pulling up to 121 mph (194 kph) from its 122 bhp engine. The car made the 0–60 mph dash in 7.7 secs. The squared wedge shape became fashionable to be seen in, and good, used examples will undoubtedly attain classic and cherished status in years to come.

RIGHT
*Early MR2s had cute, cheeky looks which appealed to sportscar-starved Britons. T-Bar roof cars are the most popular, and renowned Toyota reliability is a bonus.*

# TOYOTA SUPRA TURBO

The Toyota Supra Turbo was designed as a sporting, high-quality luxury car, exuding power and high performance, whilst carrying forward Toyota's reputation for reliability. While the running gear refrains from extrovert components, the 3-litre blown engine and the all-independent double wishbone suspension, fine drive train, including a limited slip differential and impressive cockpit, all add up to a fast coupe to be respected and enjoyed.

The Supra brims with high technology, having an on-board engine management system, a catalytic converter, an optional electronically controlled automatic transmission with two settings – one for normal driving and one for power driving – and automatic air conditioning.

The Supra is a fast, pacey coupe for the executive driver who wishes to eat up motorways but arrive relaxed. With so much torque on offer, the driving experience isn't too

ABOVE

*The Toyota Supra Turbo offers powerful touring, giving car lovers plenty to shout about. Underrated and reliable, the Supras and Supra Turbos will be good cars to hold on to.*

**TECHNICAL SPECIFICATION**

| | |
|---|---|
| **MODEL** Toyota Supra Turbo | **TOTAL PRODUCTION** N/A |
| **ENGINE** 2954cc, straight 6, electronic fuel inj, DOHC, 24 valves | **NUMBERS BUILT P/A** 1,715 sold in UK (1989) |
| **POWER** 232 bhp @ 5600 rpm/253 lb/ft @ 3200 rpm | **ENGINE/CHASSIS NOS** N/A |
| **CHASSIS** Unitary steel | **SPARES AVAILABILITY** All parts available |
| **BRAKES** Ventilated discs all round, ABS | **PRICE AT LAUNCH** £22,061 ($41,000); $30,000 |
| **TOP SPEED** 153 mph (245 kph) | **OPTIONS** Colours: White, red, black, metallics at extra cost<br>Trim: Leather trim standard on Supra Turbos<br>OE Options: High level of standard equipment, no significant options |
| **ACCELERATION** 0–60 mph: 6.1 secs | |
| **PRODUCTION SPAN** 1990– | |

demanding, and with an almost race-car type suspension, the car will handle safely and securely at all times.

The hatchback shape allows practicality to mix with sporting character, and golfing equipment for instance can be carried without recourse to the folding rear seats; once these are stowed forward, the luggage area behind the front seats becomes prodigious.

Packed with the usual array of standard fittings you'd expect from a quality Japanese manufacturer, the Supra doesn't disappoint. There are electrically powered windows, mirrors and sunroof, central locking, leather seats, alloy wheels and ABS. Cruise control is another relaxing, built-in feature. The Supra, however, is by no means a soft car; it will sprint to 60 mph in a shade over 6.0 secs, leaving much of its competition trailing.

The Supra really is edging its way into Porsche 944 territory, especially in the turbo guise. Price differentials make the Supra appear indecently cheap, or looking at it another way, incredibly good value for money. As sporting coupes go, the Supra is a quiet, refined, fast machine, one which has a loyal band of followers.

For those keen to make a statement, Toyota can help out with a special 'white pack'. This gives you a completely white Supra; the body is painted white all over, as are the exterior trimmings and alloy wheels. The effect is stunning. These are *the* Supra Turbos to buy and keep.

The Supra Turbo won't take long to force its way onto the classic scene. Already Japan is readying the next generation of top-flight coupes and the Supra will be discontinued. Reliability records are among the highest in the land, so don't be deterred by high-mileage cars.

PAST CLASSICS

## TOYOTA 2000GT

Toyota's most infamous supercar has to be the 2000GT, launched in 1966. Although only 337 of the all-aluminium bodied, backbone-chassised cars were sold, the handful of stunning convertibles are the most famous, having had a starring role carrying James Bond in the film *You Only Live Twice*.

The cars were powered by a Yamaha 2-litre straight-six engine, and produced a healthy 150 bhp. Twin wishbones and coils at the wheels gave supercar-type handling with disc brakes to ensure safe stopping. Priced at the equivalent of over £10,000 ($18,500) at launch, today the cars are worth in the region of £50,000 ($92,600) if you can find a seller.

The 2000GT never saw volume production, which is a shame as it exudes charm and character, something Japanese cars typically lacked for years. Although the 2000GT was styled on Jaguar E-type lines, it was nevertheless bravely individual.

RIGHT
*Toyota's first significant foray into the field of fast, special sportscars hardly reached fruition before the project was axed. The one-off drop head was made famous in a Bond film.*

# ALPINE-RENAULT GTA/TURBO

The Alpine company hails from Dieppe, northern France. Here during the late 1950s a racing enthusiast, Jean Redele, brought some life to stodgy Renaults by removing the engines, and placing them in steel-boned, fibreglass-bodied, rear-engined sportscars.

Alpine survived into the modern age with paternal regard from Renault, and a series of competition cars, the A108 and A110s, arrived and departed. Then in 1971 a totally new, modern shape appeared. Looking like a cast-off from an Italian design studio, aping the fabulous Ferrari Daytona nose and Maserati/Lamborghini rear end, the A310 strutted onto the scene.

By 1976 the Renault V6 engine was plumbed behind the driving seat and the real Alpine performer was born. By then Redele had lost so much money that Renault bought him out of his misery, whilst maintaining Alpine's individuality. 1985 saw an updated shell, stylish and modern, and this car continues today in normally aspirated and turbocharged forms.

Aimed at the young, sporting, affluent class of motorist, the GTA V6 has plenty of power up its sleeve. The turbo version, running an engine borrowed from the Renault 25, will see the far side of 150 mph (241 kph). Transmission runs through a five-speed gearbox, and the rear engine is mounted longitudinally, these days an unusual layout for a rear-engined car.

The fibreglass body makes up a mere 20 per cent of the total weight, most of which is attributed to the heavy steel chassis. Independent suspension all round helps keep the tail-heavy sportscar clinging onto the bends, although care is needed during heavy cornering in order to avoid a severe spin.

The GTA V6 has all the attributes of a classic supercar, rear-engine layout, great looks, pokey motor and limited build number. Despite being very much a Renault product, the car certainly isn't mass produced; marketing strategy care of parent company Chrysler shut the door on GTA V6 sales in the USA, so curtailing build numbers and ensuring a degree of exclusivity, though 20 or so cars have slipped through.

A plush leather interior is slightly marred by the corporate look of the dash, being far less sporty or attractive than the first series A310s. Nevertheless sweeping lines, radical, rakish screen angles and blistered wheelarches, make the GTA V6 a masterful and meaningful design. Never offered as an automatic, the GTA V6 is a rewarding car to drive and to look at. This is truly one of the few modern cars which looks fast while standing still.

BELOW LEFT
*The sleek Alpine wins on looks, and turbo cars are extremely fast. Sales to the US were problematic, but some cars did get through.*

## TECHNICAL SPECIFICATION

| | |
|---|---|
| **MODEL** Alpine Renault V6 GTA | **TOTAL PRODUCTION** 3,853 to end 1989 |
| **ENGINE** 2849cc, V6/2458cc V6 turbo, DOHC | **NUMBERS BUILT P/A** c.1,000 (150–200 to UK) |
| **POWER** 160/200 bhp @ 5750 rpm/166/214 lb/ft @ 2500 rpm | **ENGINE/CHASSIS NOS** N/A |
| **CHASSIS** Steel backbone/fibreglass panels | **SPARES AVAILABILITY** All parts available |
| **BRAKES** Ventilated discs all round | **PRICE AT LAUNCH** £21,620/£26,990 ($40,000/$50,000); not officially exported to North America |
| **TOP SPEED** 155 mph (248 kph) (turbo) | **OPTIONS** Colours: Standard Renault shades<br>Trim: Leather<br>OE Options: Air conditioning, alloy wheels, quality radio system |
| **ACCELERATION** 0–60 mph: 7.0/6.3 secs | |
| **PRODUCTION SPAN** 1985– | |

## SPECIAL EDITIONS

### ALPINE GTA V6 LE MANS

The summer of 1990 brought a wild-looking special edition GTA, the Le Mans from the French manufacturer. Sporting a new, even more aggressive body shape with flared wheelarches and more rounded frontal styling, over 300 are destined to be built. For British *aficionados*, there'll only be 35 to buy, all of which will be finished in 'Rouge Imperiale Metallique' with a black roof and tinted windows.

For the first time the GTA V6 incorporates a full three-way catalyst and runs on unleaded fuel, both of which restrict the performance of the turbocharged engine. Down in power to 185 bhp, the GTA Le Mans will still manage 150 mph (241 kph) and reach 60 mph in 6.62 secs. ABS is also standard as is a high-class sound system and a trip computer in the cabin. Infra-red central locking and electric windows complete the most desirable package. For all these goodies the price however, is hiked to a wopping £36,995 ($68,500).

RIGHT
*1990 saw a Le Mans special edition, complete with special metallic paint and striking alloy wheels. The Le Mans will manage 150 mph (241 kph) from its turbocharged 2.5-litre engine.*

# CITROEN 2CV

The Citroen 2CV, or Deux Cheveaux, or tin snail, as it's commonly known, is the longest-running and best-loved French car ever built. Designed before the war as a minimalist means of transport for French farmers, it had to lug potatoes and passengers across ploughed fields without upsetting their demeanour whilst merely sipping petrol, managing 70 mpg (112 kpg). The car only reached production after the ending of hostilities in the late 1940s.

Early versions were so basic as to be little more than a motorized shed, started by a lawnmower-type pull cord. Later versions gained two headlamps, a starter motor, four seats, and of course the full-length sunroof.

The 2CV only officially made it to Britain in 1974. However, from 1953–58 the 2CV was built at Citroen's Slough assembly operation in Britain mainly from parts sent from France. The venture though was not a great success.

Power, if you can call 29 bhp power, was derived from a horizontally opposed, twin-cylinder, 602cc air-cooled engine by 1974, but earlier cars ran even less potent 375cc, 425cc, and 435cc engines respectively.

Suspension was by coil springs all round, and the links were connected. Seeing a 2CV cornering hard is a dramatic experience; the body roll is spectacular, but safe. Enthusiastic drivers can virtually grind the door handles on the ground while executing sharp turning manoeuvres. Grip from spindly tyres was sufficient, and aided by the lacklustre performance.

Because of the low power output and efficient use of interior space, the car became synonymous with environmentally-conscious people during the 1970s and 1980s. Low exhaust emissions and long-lasting design meant the car would go on and on, and not need frequent replacement.

Deux Cheveaux production was halted in France in 1987, when space was needed for the new Citroen AX. Manufacture was transferred to the French auto-giant's Portuguese facility in Mangualde. Here it was built until the AX revolution surged into Portugal, causing the demise of the 2CV in the summer of 1990. Collectors and enthusiasts scrambled to buy up the last few thousand examples to preserve for posterity.

The 2CV managed to cultivate an image similar to the British Mini, by becoming something of a cult car among young, trendy Parisians. The

LEFT

*Citroen's 2CV lived for over 40 years and was made all over the world. Characterful and quirky, the 2CV offers economic driving and snail-like performance, though later Portuguese built cars could manage 90 mph (144 kph).*

| TECHNICAL SPECIFICATION | |
|---|---|
| **MODEL** Citroen 2CV | **TOTAL PRODUCTION** Seven million units |
| **ENGINE** 602cc, 2 cyls | **NUMBERS BUILT P/A** c.130,000 |
| **POWER** 29 bhp @ 5750 rpm/39nm @ 3500 rpm | **ENGINE/CHASSIS NOS** UK-built cars AZ 561000 (1956) UK-imported (1974-on) 23 KA 8001– |
| **CHASSIS** Steel platform/steel body, fabric roof | |
| **BRAKES** Discs/drums | **SPARES AVAILABILITY** All parts for later vehicles available |
| **TOP SPEED** 70 mph (112 kph) | **PRICE AT LAUNCH** c.£1,200 ($2,220) (in 1974 in UK); NA |
| **ACCELERATION** 0–60 mph: 33.5 secs | **OPTIONS** Colours: Many different colour schemes, first cars all grey. Trim: Basic cloth, though may years of vinyl OE Options: Radio |
| **PRODUCTION SPAN** 1949–1990 | |

full-length, fold-back fabric roof was a boon in the summer, and a strong selling point. The car was built on a platform chassis and was incredibly strong, despite looking as if it might collapse if you slammed the flimsy doors too hard.

When you decide to buy a 2CV, rust is going to be your major enemy, especially in newer cars. The Portuguese-built later cars used inferior metal it seems; Citroen has had to replace complete chassis on cars still under warranty due to rust. Wheelarches and most major panels are easily dismantled, so repair is basic and straightforward. Later, lighter cars will be the fastest 2CVs around.

Earlier cars are more interesting, especially those with the corrugated side panels. Many utility derivatives still exist in French farmyards. A great fun car, but for town use only.

SPECIAL EDITIONS

## DYANE/007 SPECIAL/2CV CHARLESTON

During its lifetime the 2CV sprouted many sisters, brothers and cousins, most popular of which was the Dyane. Available from 1965 to 1985, the Dyane was equipped with similar-sized or larger engines, had a squarer body and was far more adequately fitted out for modern motoring. Despite all the improvements it still fell apart from rust though.

To capitalize on the publicity from the Bond film *For Your Eyes Only*, 300 bright yellow 2CVs were sold complete with transfer kits of stick-on bullet holes and self adhesive 007 stickers. Sold at the standard price in 1981 of £2,426 ($4,500) they have to be some of the most fun 2CVs to collect.

Also in the 1980s the 'Charleston' appeared, complete with 1930s-style paint jobs in maroon and black, or yellow and black, and a higher level of luxury inside, principally soft cloth trim from 1982. In the last few years before its demise, the 'Dolly' 2CV versions were two-tone painted, up-spec'd and completed with bright green or red wheelarches and bootlids.

RIGHT
*Citroen extended the life of the 2CV with special editions like the Charleston pictured here and the Dolly. These are the nicest 2CVs to buy.*

# TALBOT-MATRA MURENA

Talbot-Matra Murena was the successor to the underpowered but pretty Bagheera, first shown to the public in 1973. Styled to incorporate three seats across, a central steering column was initially suggested, but practicality overruled its implementation.

The Murena appeared in 1980, complete with a choice of 1600cc or 2200cc engines care of PSA, the French car giant. The extra power was badly needed, though the mid-engined layout was unchanged. So too was the body-chassis combination of polystyrene plastic body, pop-riveted, bolted and bonded to the steel chassis. Panels don't craze or rust, but underneath, the metal inner panels rot gladly. Murenas generally look tidy on the outside, but under the skin they are known to be sieve-like.

The most aerodynamic, mid-engined car in the world (Cd 0.32) until the arrival of the Panther Solo, the Murena slipped quietly through the air. It was also surprisingly rattle-free and solidly built. The Talbot Tagaora-supplied engine pulled the car to 60 mph in 9.4 secs. Top speed was a little over 120 mph (192 kph).

The Murena is a far more attractive rendering of the three-across body than its predecessor, the Bagheera. Complete with pop-up headlamps, useful for judging the position of the car's nose, the steeply raked windscreen gives a supercar look. Fastback styling too gives the impression of speed, though in reality the Murena's performance was second to looks.

Dogged throughout its life by political woes, the Murena died prematurely on 1 January, 1984. It was killed off by a Renault take-over, which decreed that production facilities were needed for the new Espace model, and that there was too much competition with the existing Renault Fuego and Alpine A310 models.

When considering a Murena for your collection, bear in mind that the under chassis is vulnerable to rot. The chassis rails and sills rust easily, and are difficult to fix. Removing and replacing plastic body panels takes time and care, but they can generally be reused.

Mechanically the Murena should provide few headaches, apart from a tendency to suffer timing-chain rattle. Both headlamps should pop-up when requested, though the engine has to be running in order to provide a vacuum if you want to flash the lights often. The starter motor needs about half a day's suspension dismantling to remove, but otherwise the car is quite DIY-friendly.

Few Murenas were converted to right-hand drive, so if you do find one it's likely to be expensive. With this in mind, the best place to look for a Murena in tidy condition is in France. Prices are starting to rise for excellent examples, with pristine Murenas now worth well over £10,000 ($18,500).

BELOW
*Matra's Murena was the 1980s development of the three abreast Bagheera. Murena's fibreglass outer panels survive well, but underneath, the rust makes merry.*

LEFT

*The original Bagheera was to have the driver seated in the middle, however, luckily the Murena did not emulate this.*

## TECHNICAL SPECIFICATION

| | |
|---|---|
| **MODEL** Talbot-Matra Murena | **TOTAL PRODUCTION** 10,613 |
| **ENGINE** 2,155cc, 4 cyl, SOHC | **NUMBERS BUILT P/A** 1296,6141,2171,1004,1 |
| **POWER** 118 bhp @ 5800 rpm/139 lb/ft @ 3200 rpm | **ENGINE/CHASSIS NOS** N/A |
| **CHASSIS** Steel chassis/polystyrene composite body | **SPARES AVAILABILITY** France's PSA motor corporation parts bin provided most bits, so availability good |
| **BRAKES** Discs all round | |
| **TOP SPEED** 124 mph (188 kph) | **PRICE AT LAUNCH** c.£8,000 ($14,800); not officially exported to North America |
| **ACCELERATION** 0–60 mph: 9.2 secs | **OPTIONS** Colours: Usual Talbot shades, early Murenas finished in metallic gold. Trim: Velour seat coverings |
| **PRODUCTION SPAN** 1980–1984 | |

## SPECIAL EDITIONS

## 142 PREPARATION MODEL

Adding Weber twin-choke carburettors, a bigger bore exhaust, and a spoiler heralded the introduction of a very sporty Murena indeed in June 1982. The 142 Preparation only survived thirteen months until it became the less exciting sounding Murena S. Certainly the most desirable Murena available, the 142 was usually finished all in white, which enhanced the car's dramatic looks. Few were ever made, and they're rare outside of France.

ABOVE

*The special 142 Preparation model is one of the nicest ever Matras, but is also very rare and expensive to fix if badly rusted.*

# PEUGEOT 205 GTI 1.9

Following hard on the heels of the pace-setting Golf GTi, the car which started the hot-hatch revolution back in the mid-1970s, comes the Peugeot 205 GTi. Based around the standard-setting 205 body, the GTi with 1.9-litre engine has the power and style of a sportscar, in the clothes of a workaday shopping car.

The small Peugeots were loved by press and public at their launch, yet few could have thought how effective they'd be on the rally circuits. The 205 won the Manufacturers' World Rally Championship in 1985 with the 205 Turbo 16, and as the sales figures show, the public wanted road-going versions of the rally giant-killers.

The 1.9 GTi is a particularly effective sportscar, fast off the line, and a joy to handle. Ride is firm, and grip from both the wide 185/55 VR15 tyres and the deep, bucket, leather-trimmed seats is prodigious. Settled into the driving seat you feel as if you were strapped into a rally car. To call it the Mini-Cooper of its day would be justified as it sings along, flattening out bends, exiting corners faster than it went in, and giving the driver with speed and finesse in his heart something to smile about.

Aimed right at the centre of the hot-hatch market, the 205 GTi 1.9 provides plenty to rave about from the driver's seat, but with so much room taken up by the larger sports seats, there's barely enough space for a couple of small children in the back. The usefully sized hatch doesn't give generous loadspace unless the rear seats are lowered.

The standard car copes admirably with undulating surfaces and twisty, badly laid tarmac bends and surfaces, making country road driving a well-balanced pleasure.

Spares are easily accessed through the vast dealer network, and the car can bask in its competition image for some time to come. Unhappily the Group B rally cars were killed off in their prime, being more dangerous than they should have been. The nearest thing to a thoroughbred rally car which can handle domestic duties has to be the diminutive 205 GTi 1.9 Peugeot.

Prices, even for early cars, are still strong, held aloft by demand. This is the quintessential high-powered small, racy car so check carefully for accident damage to the shell. Signs of renewed front or rear sections could indicate trouble.

LEFT
*Peugeot's baby hot-hatch took Europe by storm, won rallies when off-road and drivers' hearts when on the street. It was slated to follow in the Mini-Cooper's footsteps as a tuned-up version of a car, perfect for its time.*

## TECHNICAL SPECIFICATION

| | |
|---|---|
| **MODEL** Peugeot 205 GTi 1.9 | **TOTAL PRODUCTION** N/A |
| **ENGINE** 1905cc, alloy block & head, fuel injection | **NUMBERS BUILT P/A** 61 (1986), 4,098, 5,512, 4,753 (1989) UK |
| **POWER** 130 bhp @ 6000 rpm/121 lb/ft @ 4750 rpm | **ENGINE/CHASSIS NOS** 07595005-(23569134-catalyst from 1990) |
| **CHASSIS** Unitary steel | **SPARES AVAILABILITY** All parts available |
| **BRAKES** Ventilated discs/discs | **PRICE AT LAUNCH** £9,295 ($17,000) UK; not officially exported to North America<br>**OPTIONS** Colours: White, red, black<br>Trim: Cloth with leather trim<br>OE Options: Power steering (1990), electric windows, central locking, leather seats, alloy wheels, sunroof, special edition with metallic green or blue paint and leather int (1990) |
| **TOP SPEED** 127 mph (203 kph) | |
| **ACCELERATION** 0–60 mph: 7.8 secs | |
| **PRODUCTION SPAN** 1986– | |

## SPECIAL EDITIONS

# PEUGEOT 205 TURBO 16

In 1984 a limited number of 205 Turbo 16s were built for the road. This followed the launch of the similarly specified, but more highly powered rally car, featuring a 16 valve engine, variable torque split, four-wheel drive and mid-mounted 1775cc engine. Similar to the Renault 5 Turbo in concept, the cars had 200 bhp on tap at 6500 rpm. The KKK turbocharger rushed this car to 60 mph (96 kph) in 5.9 secs and upped the price to around £26,000 ($48,000). Turning up the boost for competition use saw up to 420 bhp in various guises. These models will not be cheap if ever offered for sale, but true investment and interest cars never are.

RIGHT
*Peugeot took motorsport by storm during the 1980s, coming from nowhere to win rallies all over Europe. The Turbo 16 is the most sought-after 205 ever and they'll never be cheap.*

# RENAULT 5 TURBO

Not to be confused with later 1980s Renault 5 Turbos, this is the mid-engined car, of which barely enough were produced to satisfy the homologation rules. Renault designed this wide-wheeled wonder for Group Four competition use, but were forced to sell cars in road-going form first. These less powerful cars still thrust 160 bhp to the road via exceedingly wide tyres at the back, which were covered by extravagant wheelarches, themselves carrying cooling ducts for the engine bay.

Although the cars were designed in-house at Renault, the final styling was done in the house of Bertone. There were two versions of the Turbo 5: the first was a super special car, having a lavishly equipped interior and a dazzling array of instruments; the Turbo 2s had stock R5 Gordini seats and dash. Both cars used alloy roof and door panels, and glassfibre wings and arches. The limited Turbo 1s cost £15–18,000 ($28–33,000) and the Turbo 2s around £8,500 ($15,700).

You can spot the earlier T1s by the crafted tool chest behind the seats and the luggage straps atop the engine compartment behind the seats. Between 1980 and 1981 1,820 Turbo 1s were built. Turbo 2s were devised to capitalize on the demand and rally success, and a total of 3,167 left the Dieppe factory between 1981 and 1984.

The prescription for building the Renault 5 Turbo was to take a stock Gordini shell and remove the rear section, grafting in a cradle for the power unit and a frame to carry the fibreglass wheelarches. The result was a semi-space frame at the back end.

Double wishbones held each wheel to the body, mid-engine layout gave an almost ideal 45/55 weight distribution, and ventilated disc brakes slowed the snarling monster. Because the 1397cc engine lay across the back, no rear passengers were allowed. Fuel injection was taken care of by Bosch's K-Jectronic, and the Turbo ran an intercooler and reached 12.2 psi at peak pressure.

Only 20 or 30 T1s came to the UK, but there are reckoned to be almost 150 T2s here, all independently imported. Later T2s could be tuned by three methods: to 185 bhp with a bigger intercooler; to 210 bhp with the same plus polished heads, and to 245 bhp with all of the above and a strengthened bottom end.

ABOVE

*Renault's superfast Turbo cars shouldn't be confused with the standard front engine front wheel-drive mass produced cars. The specially built mid-engined cars cost almost ten times as much as their lesser siblings today.*

## TECHNICAL SPECIFICATION

| | |
|---|---|
| **MODEL** Renault 5 Turbo 1/2 | **TOTAL PRODUCTION** 1,820/3,167 |
| **ENGINE** 1397cc, fuel injection, turbo | **NUMBERS BUILT P/A** c.1,000 |
| **POWER** 160 bhp @ 6000 rpm/155 lb/ft @ 3250 rpm | **ENGINE/CHASSIS NOS** N/A |
| **CHASSIS** Unitary steel/aluminium panels and glassfibre arches | **SPARES AVAILABILITY** Reasonable |
| **BRAKES** Ventilated discs all round | **PRICE AT LAUNCH** £15,000/£8,500 ($28,000/$15,725); not officially exported to North America |
| **TOP SPEED** 124 mph (198 kph) | **OPTIONS** Colours: Red, black, metallic white, blue, bronze<br>Trim: Red leather (T1s) cloth interior later cars<br>OE Options: Tune up kits to 245 bhp |
| **ACCELERATION** 0–60 mph: 6.0 secs | |
| **PRODUCTION SPAN** 1980–1984 | |

With a deep front spoiler, complete with twin fog lamps, and a bonnet scoop (almost subtle compared with the lurid rear wheelarches) the Renault is highly noticeable in traffic. Even without the word 'TURBO' spelt out in eight-inch high lettering along both doors, you'd quickly realize this wasn't an everyday car.

PAST CLASSICS

## RENAULT GORDINI (DAUPHINE)

Sold from the late 1950s all through the 1960s until the Dauphine was succeeded by the Renault 8, the hot-shot small Renault was the Dauphine Gordini. Though it only produced 40 bhp from its 850cc engine, it was respectably quick with interesting handling to match. Like all Dauphines, the Gordini versions rust to nothing, so sound examples are rarely seen, but not expensive. If you want something completely different, these rear engine cars could be the answer. The only problem, apart from difficulties with sourcing parts, will be that the rear-mounted radiator sits directly behind the rear seats; the first small car with heated seats all year round?

RIGHT
*The Gordini versions of Renault's Dauphine are certainly the ones to keep your eyes open for, but beware of extensive rust damage.*

# ALFA ROMEO GTV6

Bertone's beautifully sweeping Alfa Romeo GTV gained a powerful V6 engine in 1981, seven years after the initial wedge shape appeared. Until then the car had been running a variety of Alfa four-cylinder engines to reasonable effect, but with the coming of the V6, true performance was now available. Aimed mainly at the USA market, the engine was bored to 2.5 litres and fitted with Bosch L-Jetronic fuel injection. The GTV6 would push to 125 mph (201 kph) with 160 bhp on tap at a noisy but extremely pleasing 6000 rpm.

Under the sleeked-back sheet metal, which by 1981 was almost completely cured of terminal rust, there was a new suspension set up, different wheels and tyres, as well as beefier brakes. The transmission still sat over the rear axle, which gave good weight distribution but a poorer than expected gearchange action. Careful, familiar selection eased the initial unfriendliness of the gearbox, and the sound of that single-overhead-cam-per-bank engine breathing deeply made the chore of gear-changing bearable.

Thrilling the Alfa enthusiast with a yen for power and style, the GTV6 was a hit. Having already cultivated a significant following with the four-cylinder engines, the V6 cars were popular. In some markets, notably South Africa, where this was offered first, a bored-out 3.0-litre motor used the potential of the V6 to the full. Complete with ground-hugging front air dam, the 3.0-litre cars were very special competition-bred machines, capable of rapid acceleration and high top speed, well over 130 mph (208 kph). Few, if any were sold in the UK, though unofficial conversions are known to exist.

Alloy wheels, tinted glass, smart interiors and fussy instrument clusters are the hallmarks of this car, and the suggestive bonnet bulge intrudes into the driver's line of vision.

BELOW

*Slotting the fabulous V6 under the GTV's sharply styled bonnet was a great move on Alfa's part. Good rust-free cars are widely available, and not too pricey: an excellent investment.*

RIGHT
*Inside, the GTV6 is a driver's car – deep bucket seats and lots of dials. Rear-seat passengers don't fare so well, but there's enough space for children, so the GTV6 is an ideal family classic.*

This is one 1980s Alfa definitely worth snaring and laying up for a rainy day. Cheap to buy now, around £5,000 ($9,000) should secure a tidy rust-free, low-mileage car having had only one or two owners. The red ones are the cars to buy – most were painted this colour – though they look just as good in dark metallics as well.

Alfa purists maintain that the four-cylinder, 'chrome bumper' cars are the most attractive classic GTVs to buy, but as with all 1970s Alfas, the rust bug does take a particular delight in ravaging the shell. The later GTVs in six-cylinder guise offer smooth power, great handling, fabulous engine note and aggressive styling, despite the plastic-covered, impact-absorbing bumpers. At the price and with the parts in good supply, they're hard to beat.

## TECHNICAL SPECIFICATION

| | |
|---|---|
| **MODEL** Alfa Romeo GTV6 | **TOTAL PRODUCTION** c.20,000 |
| **ENGINE** 2492cc, V6, SOHC, fuel injection | **NUMBERS BUILT P/A** c.4,800 |
| **POWER** 160 bhp @ 6000 rpm/152 lb/ft @ 3200 rpm | **ENGINE/CHASSIS NOS** N/A |
| **CHASSIS** Unitary steel | **SPARES AVAILABILITY** All parts generally available |
| **BRAKES** Discs all round (rear inboard) | **PRICE AT LAUNCH** £9,495; $13,000 |
| **TOP SPEED** 125mph (200 kph) | **OPTIONS** Colours: Red mostly, metallics as options<br>Trim: Cloth<br>OE Options: Air conditioning (US spec) |
| **ACCELERATION** 0–60 mph: 8.5 secs | |
| **PRODUCTION SPAN** 1981–1986 | |

PAST CLASSICS

## ALFETTA SALOON

The GTV grew out of the Alfetta saloon, a bland-looking four-door with a penchant for rust. Early GTVs were badged as Alfetta GTs. They are not yet revered as classics, but are definitely a sound option if you can find a car in tidy shape. A variety of engines is offered, from 1500cc to 1962cc, all with the delicious twin overhead cam motor. Gearboxes are mounted over the rear axle to give better weight distribution. Built from 1972 to 1984, they're good parts sources for the GTVs presently, but classics in their own time for sure.

RIGHT
*Alfa turned this average-looking saloon into a great car with the twin overhead cam engine, but the rear-mounted gearbox was never the success it could have been. The Alfetta sired the GTV, and in its own right is now becoming something of a classic.*

ITALY

# ALFA ROMEO SZ COUPE

Oh Alfa what *have* you done? Or should that be, Signor Zagato what have you done? Not content to build, like every other volume European manufacturer, a coupe based on the running gear from a top-selling run-of-the-mill box, Alfa commissioned design maestro, Zagato, to build a coupe which would be instantly recognizable. It also had to make you stop and salivate. And presto – the Sportiva Zagato.

You'll not mistake it for an Audi, Vauxhall/ Opel, Volkswagen, or Mercedes small-volume build, executive coupe. It is not a super sleek, Cd-bashing, 1990s bland styling example you'll trip over all around Europe. You'll know it instantly, and so far, with only a handful traversing the autoroutes, it's done more for Alfa's and Zagato's reputation than a whole fleet of Alfa 164 two-door coupes could with a decade of race-track effort.

Powered by a similarly sweetly desirable V6 3.0-litre engine, which sleeps under the bonnet of the stock Alfa 164, the SZ runs wild to 155 mph (290 kph). It accelerates to 60 mph around the 7.0 secs mark, and it is very low to the ground, courtesy of deep skirts and spoilers, but it isn't a low, rakish coupe. Rather it sits up, and needs to be hauled higher on its suspension, for negotiating ramps and uneven surfaces.

The SZ shares the same floorpan and suspension as the Alfa 75, including a De Dion rear axle. Height-adjustable Koni dampers are all part of the package.

The body though is something special, a plastic composite called Modar, high-tech and highly priced, but light and strong. The roof panel is aluminium and the chassis a strong steel affair. The rear Devil's collar spoiler is made from carbon fibre.

Despite the stubby appearance, the computer-aided design engineers have managed a Cd of less than 0.30. Nevertheless the top speed and performance of the V6 3.0-litre engine haven't aroused typical journalistic euphoria. A V10 engine is already well on the way, and if this finds its way into the SZ's nose, we'll have something to shout about. Those now waving £100,000 ($185,000) cheques at Alfa dealers trying to secure a current SZ would do well to hold onto their money until the vaunted V10 arrives.

Even before the first of the limited issue 1,000 cars began production, Alfa had pre-sold them at £40,000 (c.$74,000) each. Only 100 are destined for the UK; the lucky buyers are still in the queue. The American market should receive a more generous quota. Several speculators ordered cars, though in late 1990 the market is less buoyant than it was at Motorfair '89 where the car made its UK debut.

The only note of warning in this otherwise blemish-free raid on the hearts of Alfa enthusiasts and great car lovers everywhere, is that, at the time of writing, very few bluff-fronted, six-headlamp inlaid snouts have been seen shuffling their way down the special production line at Snr Zagato's place. If you're lucky enough to be in the queue, patience is still needed. If you're not, keep saving; they've got to be for sale at some point, haven't they?

RIGHT
*The SZ took the world by storm and its outrageous looks stopped enthusiasts in their tracks. Despite the pacey shape, the engine is only the 3-litre V6. Rumours of the Alfa V10 engine persist but for the time being remain just that.*

BELOW
*From the rear the SZ looks like no other car, with the high rear deck and discrete spoiler. The door wedge shape is accentuated by the rising waistline.*

## TECHNICAL SPECIFICATION

| | |
|---|---|
| **MODEL** Alfa Romeo SZ Coupe | **TOTAL PRODUCTION** 1,000 |
| **ENGINE** 2959cc, V6, electronic fuel injection | **NUMBERS BUILT P/A** c.200 |
| **POWER** 210 bhp @ 6200 rpm/181 lb/ft @ 4500 rpm | **ENGINE/CHASSIS NOS** N/A |
| **CHASSIS** Unitary steel/composite body | **SPARES AVAILABILITY** Dubious |
| **BRAKES** Ventilated discs all round, inboard at rear | **PRICE AT LAUNCH** £40,000 ($74,000); not officially exported to North America |
| **TOP SPEED** 153mph (245 kph) | **OPTIONS** Colours: Red only<br>Trim: Tan leather<br>OE Options: Interior adjustable ride height |
| **ACCELERATION** 0–60 mph: 7.0 secs | |
| **PRODUCTION SPAN** 1989– | |

## PAST CLASSICS

### ALFA GIULIETTA SPRINT SPECIALE 1960

Built by Zagato around the mechanicals of the Giulietta Sprint Veloces, the 1960 SZ had absolutely stunning proportions. The fared in headlamps and aerodynamic-inspired styling created the impression of speed even when stationary. The engines were 100 bhp @ 6000 rpm units which spirited the SZ coupes along to 130 mph (209 kph) plus. Today such cars are rarely offered for sale, but command stiff prices when they do appear. Even standard Alfa-built Giulietta Sprints can sell for as much as £12,000 ($22,000), so the SZs will be far more expensive.

RIGHT
*The name Zagato lends status and value to any car and especially to one of such beautifully aerodynamic proportions as the Alfa Giulietta Sprint Speciale.*

# ALFA ROMEO SPYDER (1990)

Having a good four years on their hands until Pininfarina comes up with the next Alfa Romeo Spyder, the Italian car maker decided to bring up to date the sadly neglected, mid-1960s Spyder for 1990. It's the car Dustin Hoffman made famous in *The Graduate*, and now in its third set of clothes, it just manages to hide its age.

After the early 'Duetto' round-tailed cars of the 1960s, the 1970s saw a Kamm-tailed version, latterly horribly bespoilered. 1990 saw a slightly more tasteful restyle; more aerodynamic and nicer with it, the rear end of the car received Alfa 164 tail-end treatment.

The beautiful twin-cam Alfa 2.0-litre engine doesn't over-exert itself to push out 118 bhp. However, if you listen to the noises even at high revs, you'll appreciate the days when the roof can be lowered by a flick of the wrist, and the sound from under the drooping curved bonnet fill the air.

The interior hasn't been smartened up by any noticeable degree, still being very 1960s with the vinyl seat coverings and big, twin dial instrument cluster. Sideways on, the new moulded bumpers and slight side skirts look too bulky for what was once a skinny and slinky shape. Nonetheless, the car is now far more aesthetically satisfying than the US-specified, rubber bumper models of recent times.

Driving a Spyder revives that summer feeling. It's definitely to be driven with the top down. Although the non-optional catalyst stifles power somewhat, the Spyder is no blood-and-guts racer, rather a relaxed open-top, sporty car for those who want to make eyes swerve in their direction.

Thin on the ground in the UK, with all official Alfa imports being delivered in left-hand drive, you'll have a hard task searching for a new Spyder. However, buying an older Kamm-tail car is quite easy; over 30,000 2.0-litre cars were built, and though many are to be found in the USA, several hundred made their way to the UK. Buyers need to beware of rust and sloppy running gear, though parts are in good supply, if expensive.

The 1990 versions are destined for classic status, since they'll be the last of a very long line of successful and image-packed sportscars. They'll also be the most expensive, though Alfa Romeo specialists now demand over £10,000 ($18,500) for even the most mundane and numerous examples carrying squared-off tails. The most charismatic cars are the round-tail early 1,300 Duettos. Though underpowered, they are delightfully nimble and balanced. The 1750cc twin cams are faster, but 2.0-litre cars remain the better bet. The last ever cars in 1993 will be collectors' items for sure.

LEFT
*A new rear light cluster and tail treatments are in line with the current corporate look. With the value of used Spyders rising, a brand new car makes more sense now.*

## TECHNICAL SPECIFICATION

| | |
|---|---|
| **MODEL** 1990 Alfa Romeo Spyder | **TOTAL PRODUCTION** More than 30,000 all types |
| **ENGINE** 1,962cc, DOHC, 3-way catalyst, elec. fuel injection | **NUMBERS BUILT P/A** N/A |
| **POWER** 120 bhp @ 5,800 rpm/118 lb/ft @ 4,200 rpm | **ENGINE/CHASSIS NOS** N/A |
| **CHASSIS** Unitary steel | **SPARES AVAILABILITY** All parts available |
| **BRAKES** Discs/discs | **PRICE AT LAUNCH** £15,950 (c.$29,500); $20,900 (base model) |
| **TOP SPEED** 118mph (189 kph) | **OPTIONS** Colours: Usual Alfa shades<br>Trim: Vinyl<br>OE Options: Power steering, radio extra cost |
| **ACCELERATION** 0–60 mph: 9.4 secs | |
| **PRODUCTION SPAN** 1990 (original design from 1966) | |

RIGHT
*Refreshed yet again, the 1990 Spyder shows how reluctant Alfa is to kill off its most famous open car. Revised frontal and rear styling removed the ugly rubber bumpers and restored some of the graceful beauty to a car made famous over 20 years earlier.*

PAST CLASSICS

## ALFA DUETTO

The 1990 Spyder has its roots in the 1966 Duetto, a boat-tailed convertible with fared-in headlamps set into a rounded nose. The Duetto, which later became known simply as the Alfa Spyder shared a range of engines, most notably the 1779cc '1750' twin camshaft engine. Built by Pininfarina, the early 1960s Duettos won world-wide appeal after starring in the film *The Graduate*. Early 1600cc Duettos sold only 6,300 examples between 1966 and 1967, but later 1750cc cars sold almost 9,000 units to 1971. Two litre cars appeared in 1970.

Early 1600 or 1750 cars in pristine condition now fetch almost £15,000 ($28,000), though with the new car still in production prices may stay depressed in the shorter term. Despite this, the Duetto is quite a good bet as a long term classic and great fun in the meantime. It is also a great looking car.

RIGHT
*Although the continued production into the 90s of the new Duetto may depress prices for the present, early 1600 or 1750 cars in concours condition still fetch up to £15,000 ($28,000).*

# DE TOMASO PANTERA GT5-S

Late in the 1960s Ford USA decided a mid-engined, two-seater sports supercar would fit quite nicely among their bread-and-butter ranges. So they came to an agreement with De Tomaso, whereby the small Italian manufacturer of impressive-looking sportscars sent over 5,000 cars to the USA in four years.

The arrangement wasn't all smiles though. After a short time, Ford saw the oil crisis killing off 10 mpg (16 kpg) cars, and rusty monocoque chassis were acutely embarrassing. They terminated the agreement, and the Pantera was destined for a quiet life after the initial flurry. It survives today, turning over at a mere 200 units a year.

In GT5-S form it sports a 5.7-litre Ford V8, capable of propelling the fat, wide car to almost 160 mph (257 kph) and 60 mph in 5.5 secs. Handling is safe and secure, due to fabulously wide, sticky tyres inside squared-off wheelarches, as well as wishbones all round, aided by coil springs.

Inside though, the Pantera is a 1970s disaster. You can't see the vital parts of the instruments, the pedals are hard to manipulate and the array of switches tumbled across the dash are hard to locate and operate. Given all these apparent woes, sales of the car in the UK are acceptable at around 15 per year.

Driving satisfaction, say the testers, is prodigious, with the thundering V8 rumbling your mind as you whistle along the straight and not so narrow.

The price of even the top specification car will trouble you for a shade over £41,000 ($76,000). That's Porsche 911 territory rather than Lamborghini Countach. In addition, you'll be hardly likely to run into a gaggle of similarly saddled, pretentious upstarts around town; the GT5-S is none too common these days.

The one thing which could put you off the car if you're after serious credibility and driving power, is the massive rear aerofoil. Detach that and rearward vision is increased dramatically.

The Pantera is a great classic to buy now, since prices are stable, and furthermore, not influenced by investors in the City in a way which bolsters/interferes with Ferrari and Lamborghini. You can get the engine tuned by any knowledgeable Ford V8 engineer.

The GT5-S looks like a supercar, drives the same way, and handles just as well. It is devoid of serious cachet because it lacks the mystique of aluminium engines with designer labels, but don't be fooled by this lack of coquettishness. The Pantera will do what you want, turn heads in the street and leave enough of the opposition for dead with its above-average performance, although the hulls do tend to rust.

TOP AND ABOVE
*The GT5-S compromised its credibility as a pure supercar with its massive rear aerofoil, complemented by the ultra wide tyres. However, the Pantera is not expensive: a good buy if you want power and looks.*

| TECHNICAL SPECIFICATION | |
|---|---|
| **MODEL** De Tomaso Pantera GT5-S | **TOTAL PRODUCTION** c.10,000 |
| **ENGINE** 5763cc, (351cu), V8, OHV | **NUMBERS BUILT P/A** c.200 p/a currently |
| **POWER** 350 bhp @ 6000 rpm/330 lb/ft @ 2500 rpm | **ENGINE/CHASSIS NOS** N/A |
| **CHASSIS** Unitary steel | **SPARES AVAILABILITY** Engine parts stock Ford, body more problematic |
| **BRAKES** Ventilated discs/discs | |
| **TOP SPEED** c.157 mph (251 kph) | **PRICE AT LAUNCH** $10,000 in USA in 1974 |
| **ACCELERATION** 0–60 mph: 5.5 secs | **OPTIONS** Colours: Various<br>Trim: Leather<br>OE Options: Air conditioning |
| **PRODUCTION SPAN** 1970– | |

PAST CLASSICS

## DE TOMASO MANGUSTA

The car which inspired the Pantera was built from 1967 to 1972. During this time only 400 or so copies of the initial Ghia design, penned by Giugiaro, were ever produced. The Mangusta was known for its speed; it would top 150 mph (241 kpg) easily, and used a stock Ford 4727cc V8 engine. The mid-mounted engine sat in a steel backbone chassis, and that and the tight cabin were clothed in a stylish and curvy steel body which managed without bumpers. A Mangusta in good condition will fetch up to £50,000 ($93,000).

ABOVE RIGHT
*The Mangusta is a gutsy performer with great looks to boot. Stock Ford V8 engine and beautiful Italian lines make this a car to treasure.*

RIGHT
*The rear view with split window was novel, but the styling a touch over fussy for such a lithe-looking car.*

# FERRARI 288GTO

The GTO (O represents the Italian for homologation, *omologato*) arrived in 1984 and set Ferrari fans' hearts beating faster than ever before. Here was a car at last that looked as if it meant business; flared arches, snarling snout and wild-door mirrors. In fact, the overall shape was far chunkier than the production 308s and 328s, and aggression positively oozed from the tail pipes.

Under the shapely rear deck hatch lurked a V8 engine, bored out to 2855cc, which, when flexed to the limit at 7000 rpm, would hurl out 400 bhp. Stopping power was courtesy of four large discs on the wheels, and a five-speed manual helped put the power down.

You'd have to have been a very special customer to get on the order books back in 1984. The GTO was one of the first of the brief production-run cars manufacturers are now turning to in order to whet the appetite of the more affluent customers. Only 20 left-hand drive versions were officially imported into the UK.

While the 1980s GTO doesn't have the racing pedigree of the 1950s GTOs, which swept many trophies back to the Modena display cabinets, the car does have strong visual appeal and the performance to impress. Designed also to compete in Group B, the car has some pedigree of its own, though Ferrari did spend far more time and effort on their Formula One programme.

The racing heritage of the new 288GTO is further enhanced by the use of fibreglass and composite materials in the body structure, while the main chassis is steel. A new engine and gearbox was developed for the 1984 car, and despite not being a 12-cylinder effort, the resulting V8 is a masterpiece of design and performance, readily racing to 7000 rpm.

A simple prancing horse device and 'GTO' lettering applied to the rear panel announce the 288, but the give-aways are yellow Ferrari shields on the flanks, deeper spoilers on the front and along the sides, and the stand up rear view mirrors acting as flag bearers on each door.

RIGHT
*Aggressive styling and wide mirrors mounted on turrets, identify this red supercar as the 288GTO. A worthy namesake to the earlier 1960s car, though it was never given sufficient opportunity to flex its muscles on the track.*

BELOW
*Staring the 288GTO in the mouth is an awesome sight: only front on do you realise how wide it really is. Cars for sale will never be cheap.*

Top speed is around 190 mph (305 kph), with 60 mph arriving in a brisk 5.0 secs. Designed specifically for racing, the GTO never made the circuits buzz, since Group B was dismantled before the raging red of the prancing horse could make it to the chequered flag first. Some customers may have plumped for the option package of air conditioning radio, and electric windows, but real drivers take their GTOs as they come, ready for track battle rather than shopping or cinema car parks.

## TECHNICAL SPECIFICATION

| | |
|---|---|
| **MODEL** Ferrari 288 GTO | **TOTAL PRODUCTION** 273 |
| **ENGINE** 2855cc, V8, quad cam 16 valve, twin turbo | **NUMBERS BUILT P/A** 273 |
| **POWER** 400 bhp @ 7000 rpm/366 lb/ft @ 3800 rpm | **ENGINE/CHASSIS NOS** 52469-56775 |
| **CHASSIS** Space frame tubular steel chassis/Kevlar body panels | **SPARES AVAILABILITY** Ask Ferrari |
| **BRAKES** Ventilated discs all round | **PRICE AT LAUNCH** £73,000 ($135,000); c$90,000 |
| **TOP SPEED** 190 mph+ (304 kph+) | **OPTIONS** Colour: Rosso red<br>Trim: Sporty red/black leather<br>OE Options: Air conditioning, radio cassette, electric windows |
| **ACCELERATION** 0–60 mph: 5.0 secs | |
| **PRODUCTION SPAN** 1985 | |

## PAST CLASSICS

# FERRARI 250GTO

The 1960 namesake, the Ferrari 250GTO, is now one of the world's most expensive cars, auction prices reaching £10m ($18m). Mind you they are great-looking, great-sounding cars and have a highly successful racing pedigree. The front-engined 3.0-litre V12 250GTO emerged in 1962 and won the world GT championships that year and again in the following two years. Le Mans also fell to the 250GTO in 1962.

RIGHT
*Exuding power, the muscular 250GTO can now fetch almost £10m ($18.5m) at auction: one of the most valuable cars in the world.*

# FERRARI TESTAROSSA

The 1980s did not kill off the supercar as everyone had predicted. Rather the supercar builders fought back at restrictive legislation, and created beauty around regulations. Complete with striking styling, agility in handling and awesome grunt in acceleration, the new, flat 12-engined, nostalgically-named Testarossa claimed the hearts of the world's motoring *fundis* at the Paris Motor Show in 1984.

Ever since, the car has been the acme of supercars. Rounded, sleek, proportionally perfect and with side ribbing resembling overgrown speed stripes, the Testarossa is unmistakable in any setting.

Built mostly from aluminium, the Pininfarina shape is designed to limit downforce, though the Cd of 0.365 does betray the penalty of minimum lift. But who really notices? Once the beefy, flat 12 is fired up and warm, the Testarossa will sprint to 60 mph in a little more than 5 secs and will keep going to just over 170 mph (272 kph) if the road is long enough.

The motivating force is a Ferrari BB engine with a tune up. Twenty four valves get the breathing going at the top of the flat 12, and there's electronic fuel injection and twin intercoolers to further boost performance. The engine is mounted lengthways, of course, and there's a twin-plate clutch for swapping the ratios.

The most lasting impression left by the Testarossa is width – almost six and a half feet (two metres) in fact. The curious wing mirror which sat half-way up the windscreen pillar like a grotesque wart, partially spoiling the wonderfully powerful front end-on appearance, was later discarded in favour of a more successful use of reflecting glass.

On the move the 5.0-litre engine makes its presence felt, but as a result of careful siting of the cooling system, the passenger compartment doesn't resemble a Naples bakehouse on a summer's afternoon.

You'll know that the car was designed with American legislation in mind. The tyres are as wide as they might be, to give phenomenal grip and hardy cornering, and the car isn't light at 3,608 lb (1669 kg), so the top speed approaching 180 mph (289 kph) is all the more acceptable. For a market where the legal limit doesn't require the need for more than the Ferrari's first two gears, the car has been honed to provide snazzy, safe transport for a wealthy but racy clientele.

ABOVE
*The interior is stark but comfortable, mirroring the outside's restrained styling. The typical Ferrari 'wand'-type gear shifter and dished steering wheel are welcome relief from the bland seats and dash.*

Crystal ball gazers have already marked the Testarossa's card, and prices for these mid-engined supercars did flutter skywards during the mad period after the death of Enzo Ferrari. Now, with production still on stream, and the new 348tb on the shelves, the Testarossa's selling price has retarded a touch. It will always be a valued car, but never in the F40 or Daytona league because there are too many and it doesn't have a racing pedigree. Having said that, it was designed for the boulevard, not the Brickyard.

## TECHNICAL SPECIFICATION

| | |
|---|---|
| **MODEL** Ferrari Testarossa | **TOTAL PRODUCTION** c.3,000 |
| **ENGINE** 4942cc, flat 12, DOHC, 48 valves, electronic fuel inj. | **NUMBERS BUILT P/A** c.1,000 |
| **POWER** 390 bhp @ 6300 rpm/361 lb/ft @ 4500 rpm | **ENGINE/CHASSIS NOS** Engine (1985) 56141-,(1986) 60815-, (1987) 68013-, (1988) 75506-, (1989) 79765-, (1990) 83900- |
| **CHASSIS** Tubular steel | |
| **BRAKES** Ventilated discs all round | **SPARES AVAILABILITY** Excellent but expensive |
| **TOP SPEED** 171mph (274 kph) | **PRICE AT LAUNCH** £91,195 ($169,000); c$90,000 |
| **ACCELERATION** 0–60 mph: 5.2 secs | **OPTIONS** Colours: 18 colours available, including green, yellow and white: 80 per cent of cars are red. Trim: Cloth/leather OE Options: Air conditioning, electric windows, full leather trim |
| **PRODUCTION SPAN** 1985– | |

ABOVE
*Stylish and smart, the Testarossa won the hearts of the world even with its rather conservative styling. Instantly recognizable with those unsubtle side ribbings, the Testarossa could grace the most select of driveways.*

RIGHT
*The Ferrari 512BB was a big, beautiful supercar, capable of 170 mph (273 kph) and now frighteningly expensive. Every serious Ferrari collector should have one.*

PAST CLASSICS

## 365GT4BB/512BB BOXER

Mid way between the Daytona and the Testarossa lies the Boxer, for all three sported the flat 12 engine. The BB has its engine slung between the axles behind the passenger cabin, and it looks absolutely brilliant. A tubular steel chassis has monocoque elements, and the styling was again courtesy of Pininfarina. Top speed of the 360bhp models was in the region of 160 mph (257 kph) if you believed the testers, or over 170 mph (273 kph) if you take Ferrari's word. Either way, the BBs are immortal cars, very expensive and jealously guarded. In production between 1973 and 1984, approximately 2,300 365GT4BBs and 512BB Boxers were built.

# FIAT X1/9

If there's one car with so much potential that it cried out for development throughout its life but which was cruelly ignored, it must be the Fiat X1/9. Styling master Bertone presented Fiat with a great design, back in 1972, and produced a strikingly different wedge-shaped two-seater sports, complete with mid-engine layout for optimum handling.

The car wasn't an immediate success, saddled with stock 1300cc Fiat power units, but when this was beefed up to 1500cc, sales increased. The car performed adequately, but more than that it looked sharp. In production until 1989, the initial shape survived intact, Fiat seeming unwilling to spend money updating it.

Fiat knew lots about building small, nippy sportscars. The X1/9 was perfect for the man who wanted sporty driving and supercar handling. It had the benefit of a targa roof, but avoided the hassles of complex engineering or tatty build quality from a jumped-up kit manufacturer. Mid-engine layouts were, until then, largely the preserve of the supercar manufacturers, for whom Bertone had been slaving previously.

Later 1980s cars gained fuel injection and a five-speed box to keep up with modern machinery from Japan. Apart from having an inaccessible engine bay, the X1/9 benefited from its Fiat parentage by collecting all of its running gear directly from the corporate parts bin.

Replacement bits now are easy to come by and not particularly expensive to buy, except for Bertone panelling

The X1/9 was aimed at the sporty buyer, interested in style rather than raw performance. In 1500cc fuel-injected form it would reach 110 mph (176 kph), but even then it's great weight penalized top-end speed. The X1/9 was designed with US safety regulations in mind and so was built very solidly indeed.

There were luggage lockers front and rear, but with the targa roof stowed in the front, space was minimal. Interior trim was always sporty if not at the height of good fashion. This changed after Fiat ceded the manufacturing of the car to Bertone in 1981, the X1/9 benefited from several special editions which made it more attractive to style-conscious motorists.

Underrated and under-evolved, the X1/9 currently languishes in the doldrums, however in a short space of time this car will become regarded for what it truly is – a low-priced, neatly conceived, sporting package without too many pretentions to grandeur.

The key is to find one which has been well looked after, serviced regularly and is as rust-free as is possible from an Italian car. Bertone-built cars will probably have been better constructed, but as they are newer, they'll command a higher price.

BELOW

*When Bertone took over production of the X1/9 subtle changes kept the car going. However, the overall shape remained unchanged during the model's entire life.*

## TECHNICAL SPECIFICATION

| | |
|---|---|
| **MODEL** Fiat X 1/9 | **PRODUCTION SPAN** 1972–1989 |
| **ENGINE** 1290cc, SOHC/149cc SOHC | **TOTAL PRODUCTION** c.200,000 all types |
| **POWER** 75 bhp @ 6000 rpm/71 lb/ft @ 3400 rpm 86 bhp @ 6000 rpm, 86 lb/ft @ 3200 rpm | **NUMBERS BUILT P/A** c.16,000 |
| | **ENGINE/CHASSIS NOS** N/A |
| **CHASSIS** Unitary steel | **SPARES AVAILABILITY** Good genuine panels dear |
| **BRAKES** Discs/discs | **PRICE AT LAUNCH** £2,300 ($4,000); not officially exported to North America |
| **TOP SPEED** 99/110 mph (158/176 kph) | **OPTIONS** Colours: Special metallics on first cars, later standard Fiat shades, some Bertone cars two-tone. Trim: Cloth |
| **ACCELERATION** 0–60 mph: 15.3/10.8 secs | |

## PAST CLASSICS

# FIAT 850 COUPE

The grown up Fiat 850 saloon spawned several interesting off-shoots and the 850 Coupe is one of the prettiest. Still available in great numbers in its home country, the 843cc or 903cc engines gave the Coupes spritely performance for their size while handling is often raved about by Fiat fans. With almost 400,000 built between 1965 and 1973, finding an example won't take you to the ends of the earth although none was officially exported to North America. UK sales were limited somewhat. Rust though will be, as ever, the deciding factor: rusty cars just aren't worth the effort, quite yet. Prices range from £1,500 ($2,800) in good condition, although it is worthwhile noting that Fiat 850 Spyders – same car, no roof – are worth up to three times as much.

ABOVE
*At around £1,500 ($2,800) for vehicles in good condition, prices for Fiat 850 Coupes haven't reached classic levels.*

# LAMBORGHINI COUNTACH ANNIVERSARY

The name 'Countach', styled by Marcello Gandini for the 1971 Geneva Motor Show, evokes much euphoric enthusiasm.

Since then the Countach has ruled the world's supercar league in looks and, for a while, in performance. While there have been better cars in certain aspects of performance or build, the stunning appearance of the Countach has consistently been able to stop the traffic with its great looks. Some say the versions with outlandish wings on the tail, unfeasibly large tyres and flared wheelarches are too outrageous for a serious classic supercar, but it is undeniably one of the greatest of its era.

Now deposed by the Diablo, production has finally ceased. Under 2,000 Countachs were ever built, testimony perhaps to the shakey years Lamborghini traversed during the 1970s, when Countach fever was at its highest. American interest in the car waned during the mid-1970s–early 1980s due to import difficulties caused by legislation restrictions. Sales in Europe were small, but this has endowed the Countach with an air of exclusivity and rarity.

Structurally, the Countach is a multitube space frame, with fibreglass inner panels and hand-beaten alloy outers. Double, unequal-length wishbone suspension all round is aided naturally by significantly sized anti-roll bars and stopping is effected by chunky discs at each corner. There's no power assistance to the steering.

The Countach has always used a longitudinally placed mid-engine layout with the gearbox mounted ahead of the engine. The propshaft travels backwards to the rear wheels through a sealed channel in the crankcase.

The cabin is a tight fit for taller people – headroom was never a strong point – and first owners can have the interior built around them. A visit to the factory for a fitting is not unheard of; what's an extra £1,000 for a quick trip to Sant' Agata, when the car costs 100 times that? Even so, the untidy array of instruments cried out for a serious redesign for ages, but was palmed off with mild re-arrangements.

The Anniversary cars are the fond farewell which some say is overdue. Gandini warmed over the styling again, Lamborghini tuned the V12 for a bit more go, and the price jumped accordingly.

BELOW AND BOTTOM
*This is Marcello Gandini's final rendering of the instantly identifiable Countach shape. Squared arches are Anniversary trade marks but the rear jumbo jet wing has gone. Most cars sold are painted red. It looks fast even when standing still, and on the move it is one of the fastest road cars ever. The Countach Anniversary has been somewhat eclipsed by the later Diablo, but it will still live on for eternity.*

At least, the restyle of the European-specification cars manages without the Can Am wing on the engine cover and the horrendous impact-absorbing bumpers. The side skirts are a bit over the top, but it is the last bash, so it goes out in style.

With so many Anniversary cars produced in relation to total production, these should be easy to find when you're looking for a Countach. Obviously the most attractive for their run-out status, they do however come a close second only to the LP5000 'S' Quattrovalvole cars for overall desirability. Nevertheless, if you want a Countach that badly, it won't really matter which model you choose.

## TECHNICAL SPECIFICATION

| | |
|---|---|
| **MODEL** Lamborghini Countach Anniversary | **TOTAL PRODUCTION** 430 (c.2,000 all types) |
| **ENGINE** 5167cc, V12, 48 valves, DOHC, 6 Weber 44 DNCF carbs | **NUMBERS BUILT P/A** c.170 |
| **POWER** 445 bhp @ 7000 rpm/367 lb/ft @ 5200 rpm | **ENGINE/CHASSIS NOS** Last car no ZA 9a00005AOLK12080 |
| **CHASSIS** Steel/fibreglass and alloy panels | **SPARES AVAILABILITY** All parts available at present |
| **BRAKES** Ventilated discs all round | **PRICE AT LAUNCH** £98,857 ($183,000); $146,000 |
| **TOP SPEED** 183mph (293 kph) | **OPTIONS** Colours: Red, black white, blue<br>Trim: Leather<br>OE Options: Electric windows |
| **ACCELERATION** 0–60 mph: 5.0 secs | |
| **PRODUCTION SPAN** 1989–1990 | |

PAST CLASSICS

## LAMBORGHINI MIURA

Three different versions of the superbly styled Miura were built – the standard car, the Miura S and the SV. Power from the V12 engines was rated at 370 bhp, and latterly 385 bhp, for the SV. All cars could sprint past 170 mph (273 kph) and 475/140/150 examples of each type were created. The car lived for six years, between 1966 and 1972, and set Lamborghini up as true supercar makers. It was styled by Bertone as usual, with Dallara developing the chassis. Aerodynamics at the top end of the power range made the car's handling unpredictable, but the 3939cc-engined cars all looked stunning. Prices today are phenomenal, but worth the money.

RIGHT

*The Miura SV manged to produce 385 bhp while looking unfussy and generally superb. Broad-nosed but kind-hearted, the Miura is a joyful car to drive.*

# LAMBORGHINI DIABLO

It's called the 'Devil', but the looks aren't *that* devilish, since Gandini's original designs, full of their usual flair, were deemed too adventurous by the Lamborghini parent company, Chrysler. It does look swell though, capturing the image of 1990s supercar motoring, being the first to fly well past the 200 mph (321 kph) barrier. It's quite comfortably passed it in fact, getting to 202 mph (325 kph), taken all that way by the fine 5.7-litre V12, plumbed carefully behind the two seats in the crafted leather-lined cabin.

The new Diablo had its work cut out to better one of the world's finest supercars and one which still causes a stir wherever it turns up. Even after having been around for almost 20 years, the Countach can still cut the mustard. Dreaming up the Diablo recipe demanded far more torque to whistle past the 200 mph barrier, as well as to meet the increased demands on suspension, aerodynamics and body construction.

The new chassis is made of squared steel tubing, and the crumple-zone protected cabin has a steel shell hand-grafted onto the frame. Unlike the aluminium hand-crafted doors of the Countach, the Diablo gets alloy pressings, which improve fit quality. Lamborghini developed a special composite for the bumpers, sills, bonnet and engine cover. Called 'Autoclave', it's twice as light as aluminium and far lighter than steel, having greater torsional stiffness.

The running gear uses a five-speed manual, or three-speed semi-automatic (clutch pedal-free) transmission with a four-wheel drive with vicous traction system under development. As yet there's no ABS either, but nonetheless, the technical layout of the Diablo is most impressive.

Even if you're not sent into raptures by the technical specification, the thought of the svelte body and tip-up doors should grab you. Also, it is the first really important Lamborghini for years, and it is rare.

Because so many sales are to the USA, luxury equipment which has no real place in an out-and-out sports machine is included in the standard fittings list. There's air conditioning, electric windows, superior sound system, as well as the involved catalyst equipment filling the engine bay

FAR LEFT
*Inside the Diablo there's lots of sumptuous luxury, but sensible support for a driver needing to concentrate at over 200 mph (321 kph).*

BELOW
*The new Diablo shows off its butterfly doors, a development of the system used on the legendary Countach. Diablos are made at the rate of about 10 per week.*

with mufflers and pipes. The interior is uninspiring, lots of leather, impact-absorbing foam dash and undramatic, laid back instruments; the stalks are from a BL Marina for goodness sake!

While you're hanging around waiting for one – the list stretches to three years as production is still limited in the start-up phase – consider the fact that in over 15 hard, sometimes on-the-brink-of-bankruptcy years, Lamborghini only managed to turn out around 2,000 Countachs. Plans are afoot for 500 Diablos a year if Chrysler have their way.

## TECHNICAL SPECIFICATION

| | | | |
|---|---|---|---|
| **MODEL** | Lamborghini Diablo | **TOTAL PRODUCTION** | N/A |
| **ENGINE** | 5729cc, V12, 48 valves, sequential fuel injection | **NUMBERS BUILT P/A** | c.350 (500 units p/a projected) |
| **POWER** | 492 bhp @ 7000 rpm/428 lb/ft @ 5200 rpm | **ENGINE/CHASSIS NOS** | N/A |
| **CHASSIS** | Steel tubes/body; alum., carbon fibre comp., steel | **SPARES AVAILABILITY** | Good but costly |
| **BRAKES** | Ventilated discs all round | **PRICE AT LAUNCH** | £115,000 ($213,000); $226,000 (1991) |
| **TOP SPEED** | 202 mph (323 kph) | **OPTIONS** | Colours: First six months four colours only; red, yellow, black, white. Trim: Leather. OE Options: Luxury items as standard include electric windows, central locking, air conditioning |
| **ACCELERATION** | 0–60 mph: 4.1 secs | | |
| **PRODUCTION SPAN** | 1990– | | |

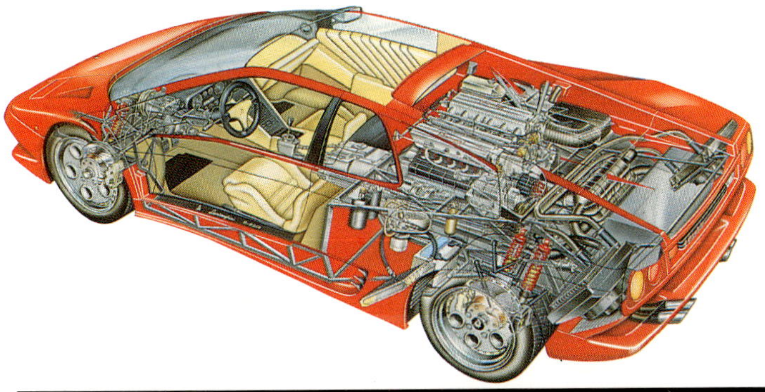

RIGHT
*The Diablo laid bare, showing chassis and engine layout. The rear-mounted longitudinal V12 engine pushes out 492 bhp.*

## PAST CLASSICS

### LAMBORGHINI 350GT

Ever ready to take on Signor Ferrari, Ferrucio Lamborghini used his tractor and air conditioning fortune to create a superior manufacturing legend. His first car, the 350GT (1964–1967), was a two-seater, powered by a quad camshaft V12 3497cc engine. Only 130 or thereabouts were built before, it was eclipsed by the arrival of 400GT in 1966, with a bigger engine and two more seats. Heralding the birth of the Lamborghini dynasty, cars like the 350GT and 400GT don't come cheap, if they are ever offered for sale, and as the first children of the Lamborghini dynasty, they'll always be sought after. Prices start at £100,000 ($185,000) for good ones.

RIGHT
*The 350GT's production life was curtailed by the introduction of the 400GT in 1966. Indeed only around 130 V12 3497cc vehicles ever hit the streets.*

I T A L Y

# LANCIA DELTA HF INTEGRALE 16V

Born and bred for competition and homologated for road use, the 16-valve Integrale is pure rally car made comfortable. Even before you slide into the contoured, gripping front recaros, the wheelarch blisters on the rear doors, the wider tyres and the red 16-valve badge tell you this is a fast but handsome brute.

Built only as a left-hand drive car, it has spearheaded Lancia's effective attempt to claim the world rally crown. In 1989 it was successful; the four-wheel drive, 16-valve car beat stiff opposition from Audi, Toyota and Ford.

The 16-valve engine also boasts counter-rotating balancer shafts, fuel injection and almost lag-free turbo boost. The transmission splits the power 47/53 per cent to front and rear respectively, giving the rallying crews a stable base from which to operate. It also means road drivers will find almost optimum handling abilities.

Inside you'll notice a lack of leather, but will find charming striped seat coverings in cloth instead. Deep bucket seats hold driver and passenger firmly, but securely for those G-Force inducing corners. The instrument cluster features yellow on black dials, which, admittedly, are not to everyone's taste.

The Lancia is a hard-working, modern sports hatch. It greedily dispenses with petrol – a gallon lasts a scant 17 miles (27 km) – but from the outside, looks far more racy than say a close rival, the Audi Quattro. Rarity in the UK will bolster prices, and I daresay interest too. Continually pulling in rally victories will encourage a strong and loyal following.

ABOVE
*Eight-valve engined Deltas were prominent before the 16V appeared, but the later car (pictured here) is the better of the two. Pre-owned cars aren't too expensive yet.*

their following grew. As rally victories continued so demand for rally replicas mushroomed.

Finding a good, well-cared for Delta will supply you with endless days of motoring fun. It's a capable car, which can rush great distances without sending you to sleep, keeping you fully

| TECHNICAL SPECIFICATION | |
|---|---|
| **MODEL** Lancia Delta HF Integrale 16V 4 | **TOTAL PRODUCTION** N/A |
| **ENGINE** 1995cc, DOHC, Garrett T3 turbocharger, fuel inj. | **NUMBERS BUILT P/A** N/A |
| **POWER** 200 bhp @ 5500 rpm/200 lb/ft @ 3000 rpm | **ENGINE/CHASSIS NOS** N/A |
| **CHASSIS** Unitary steel | **SPARES AVAILABILITY** All parts available |
| **BRAKES** Ventilated discs/discs | **PRICE AT LAUNCH** £20,350 ($38,000); not officially exported to North America |
| **TOP SPEED** 137mph (219 kph) | **OPTIONS** Colours: Predominantly red. Trim: Cloth OE Options: Central locking, trip computer, power steering, alloy wheels, electric front windows, Recaro front seats, 4-channel ABS, metallic paintat extra cost |
| **ACCELERATION** 0–60 mph: 5.7 secs | |
| **PRODUCTION SPAN** 1987– | |

The Delta is a pretty unusual car, in that for the first section of the model's life it was just another hatch from Italy. Suddenly the rally engineers got their act together, and far more exciting versions appeared – the HF, HF Turbo and finally the Integrale. As the cars got faster, briefed through feel and feedback. Keeping one in tip-top condition won't be cheap, but to reassure you, a short drive, or a long, thoughtful stare at the aggressive wheelarch flares, deep front air dam or bulging front bonnet will undoubtedly have you reaching for the keys.

ABOVE
*Lancia won rallies throughout Europe with these cars, though the early Deltas weren't anything special. Buy a used one with care, as they tend to be driven hard.*

PAST CLASSICS

## LANCIA FULVIA HF

No, HF doesn't stand for High Flyer, but it could do. The HF was the road version of the successful Lancia Rally car, and between 1966 and 1972, over 7,000 were produced in various guises, most with alloy panels, special glass in the windows and with a choice of 1.3-litre or 1.6-litre engines, tuned for the enthusiast. Faster 1.6-litre jobs would crack 110 mph (177 kph) with ease. Prices now range from £7,000 ($12,000) for cars in good order, but pristine cars are more expensive still. This is a classic with true grit, competition history and Lancia engineering.

RIGHT
*Built between 1966 and 1972, production of the Lancia Fulvia eventually ran to over 7,000. Both 1.3- and 1.6-litre engines were available.*

# LANCIA MONTECARLO

This is the car which lived twice. Launched in 1975, after being conceived as the bigger brother of the Fiat X1/9, the Montecarlo was killed off in 1978 after a spate of serious problems and industrial heartache in Italy had strangled sales.

Then, just when people had started to forget about the mid-engined two-seater, it won the World Rally Championships Manufacturers' prize, and it was back in production. Sadly the turbocharged engine was missing, but the old 120 bhp 2.0-litre twin cam was back behind the seats. In the USA emission controls led to a 1.8-litre engine which produced a miserable 84 bhp being used, and the name 'Scorpion' was stuck on its rump, since GM's Chevrolet division already had the Chev Monte Carlo.

So it was back in 1980/81 in the UK – and the rear flying buttresses were back on all cars too. Initially Lancia had significantly to re-engineer the car for the UK, since the side upper panels at the rear were solid and opaque on European-specified cars. The authorities in Britain demanded better visibility, so the buttresses appeared. In the MkII versions, all cars ran see-through glazed panels.

The front end was re-vamped slightly, and the brake servo, which ran to the front wheels only and had caused terrifying premature lock-ups on the earlier cars, was dropped. Later cars, in production until 1984, were slightly less rust-prone, though today to find an uneaten one would be a major coup.

Montecarlos love dying of tin worm, and replacement non-original body panels, though available from specialists, are exceedingly dear. Benefits of ownership however, are great handling abilities, even in the wet, a comfortable interior

BELOW

*The Montecarlo lived twice, and the second series cars are the ones to buy. With rust and braking problems sorted out, the 1980s cars represent good usuable classics.*

INSET

*The Montecarlo's speedo, calibrated to 260 kph, was probably some kind of in-joke at Lancia.*

## TECHNICAL SPECIFICATION

| | |
|---|---|
| **MODEL** Lancia Montecarlo | **TOTAL PRODUCTION** 7843/7595 |
| **ENGINE** 1995cc, DOHC | **NUMBERS BUILT P/A** 843 average |
| **POWER** 120 bhp @ 6000 rpm/126 lb/ft @ 3400 rpm | **ENGINE/CHASSIS NOS** N/A |
| **CHASSIS** Unitary steel | **SPARES AVAILABILITY** Good availability through specialists, panels expensive |
| **BRAKES** Discs/discs | |
| **TOP SPEED** 120mph (192 kph) | **PRICE AT LAUNCH** N/A; not officially exported to North America |
| **ACCELERATION** 0–60 mph: 9.0 secs | **OPTIONS** Colours: Various Fiat/Lancia shades, most white or red Trim: Cloth. OE Options: Re-introduced model (1981) had glazed inside windows and front and rear styling changes |
| **PRODUCTION SPAN** 1975–1984 | |

finished in cloth, and great sporty looks. Dress one up in those Lancia racing team colours of swervy black stripes and you'll have the crowds craning their necks as you go past.

Today you'll find Lancia Montecarlos in various columns of the 'For Sales'. Early cars are best avoided, unless you're after cheap looks. The lock-up-at-a-touch front brakes meant frequent visits to the panel beaters for cosmetic surgery, so watch out. Later cars were made of better metal too, and therefore less prone to rust. 1980s cars are obviously the ones to look for as investments, but for a remarkably low price a shabby-looking Montecarlo can be a great friend.

Several special-edition cars were built in Italy and only a few ever left the Continental mainland. The ones which did are almost all bright red in colour, and worth considerably more than the stock item. They're also generally in far superior condition.

PAST CLASSICS

## LANCIA STRATOS

Lancia are no strangers to the World Championship rally crown. During the 1970s the Stratos, a mid-engined, glassfibre-bodied homologation special collected lots of prizes. Although cramped inside, the wedge-shaped body resembles a stunted Lamborghini Countach, and the power is from a Ferrari Dino 246GT. Only 492 were ever released, and most went straight to competition teams. Original road-goers are now very valuable.

RIGHT

*Cramped inside, but with a broad windscreen giving a panoramic view, the Stratos stormed to rally victories during the 1970s. Less than 500 were ever built, but replicas are around; buy with care.*

# MASERATI SHAMAL

The Maserati Shamal is a car to shock. Based loosely around the style of the Karif and the Spyder currently in production, the Shamal has a purposeful air about it. Whether it is due to the rather mean-looking front end spattered with light lenses, or the deep front spoiler or the flared wheelarches, this is the current two-door Maserati to become excited about.

Produced in handfuls only, and first seen in the UK at the 1990 NEC Motor Show, when they were flown in and out especially for the event, the Shamal has been making headlines, and creating queues of eager Maserati fans, thirsting for a new potent machine, since its official press launch in December 1989.

Behind all the fuss is the fact that, after a Fiat buy-out of the Maserati stock, Marcello Gandini strolled along and penned a nice new flagship body for the ageing and none-too-special Biturbo range. The Karif and Spyder had done their work with pepped up specification, including four-valves-per-cylinder engines, and the lovely rag-top Spyder, but pure-blooded Maserati fans wanted more. The Shamal will deliver.

Maserati decided that it needed a new coupe to attract the very cream of the sporting market, and the price of well over £70,000 ($130,000) reflects this. Engineering innovation in the shape of the new V8 engine, which, incidentally, can be pulled out of the front of the car for servicing by unbolting a front panel, provides more than adequate performance at 325 bhp.

To differentiate the Modena factory's super sporting cars from the standard *gran turismo* models, special Maseratis have been named after the winds; the Shamal is a hot and strong wind from the plains of Mesopotamia.

The engine of course is a fine, sporting design. Not only are there 32 valves, four camshafts, alloy heads and wet cylinder liners, but the electronic management system can instruct the car to run on just four cylinders should a problem develop within the bowels of the engine.

There's a bonnet scoop to keep wind and rain from the front screen and 'A' pillars, thus serving to keep the screen clean, and prevent howling winds from disturbing the soft quietness of the interior.

LEFT
*The rear view is more pleasing. Note the odd design of the rear wheelarch, styled deliberately to be different.*

## TECHNICAL SPECIFICATION

| | |
|---|---|
| **MODEL** Maserati Shamal | **PRODUCTION SPAN** 1990– |
| **ENGINE** 3217cc, V8, twin IHI turbos, quad camshafts, 32 valves, alloy block, electronic fuel injection, catalyst | **TOTAL PRODUCTION** 40 (planned) |
| | **NUMBERS BUILT P/A** N/A |
| **POWER** 325 bhp @ 6000 rpm/320 lb/ft @ 3000 rpm | **ENGINE/CHASSIS NOS** N/A |
| **CHASSIS** Maserati Spyder's steel floorplan/steel monocoque | **SPARES AVAILABILITY** Similar availability to biturbo range |
| **BRAKES** Ventilated discs all round | **PRICE AT LAUNCH** c.£70,000 ($130,000); NA |
| **TOP SPEED** 162 mph (259 kph) | **OPTIONS** Colour: Red. Trim: Leather. OE Options: Six-speed box, four-way adjustable electronic active suspension, power steering, power seats |
| **ACCELERATION** 0–60 mph: 5.3 secs | |

ABOVE
*Styled to be noticed, the Shamal is full of curves and bumps, denoting muscularity. It is not a sweet line to be seen, but effective styling catches the eye.*

Being built for speed, the Shamal has subtle aerodynamic aids as standard, including side skirts and a deep front air dam. There's the hint of a spoiler lip at the edge of the boot lid to permit a slight increase in downforce at higher speeds and create a tidier flow through the air.

The Shamal will go straight into the classic car hit parade. It has the three classic ingredients: it's rare, beautiful and red. It's also Italian and it has a V8 engine, something car enthusiasts understand so well. In performance terms it's as hot as the name suggests, and is priced well into supercar territory, which is worth bearing in mind if you're not convinced by the latest offerings from the other motor shops down the road in Modena.

PAST CLASSICS

## MASERATI INDY

Like many Maserati road cars before it, the Indy came with a front-mounted V8 engine, bored out to 4136cc at first. Later cars ran larger 4719/4903cc engines of similar configuration, all having quad camshafts. The Indy was the first monocoque Maserati, and had plenty of style at launch in 1969. It was effectively replaced by the Khamsin in 1973. The 4.7-litre engines took the car to almost 160 mph, and all-disc brakes pulled it to rest again. The Indy was quite a good seller for Maserati, with 1136 being sold over five years. The smooth-looking Vignale shape is enduringly attractive, although the bodies tend to rust. Expect to pay £35,000 ($65,000) for a smart one.

RIGHT
*The front-engined V8 Indy now fetches a good £35,000 (c.$65,000),but this will surely increase in the future. The top speed is near enough 160 mph.*

# AVANTI

The Avanti's unique story began in the 1960s with Studebaker, an American car giant with failing health. They built the first Avantis, but when production moved to Canada, two enthusiasts, Leo Newman and Nathan Altman, secured the rights to continue making the Avanti on a small and specialized scale at the old factory in Indiana.

That was back in the mid-1960s. By the late 1970s production was still limited to 300 units a year, all to individual, and often garish, specification. Eventually both Newman and Altman died leaving the company in the hands of Stephen Blake who revitalized the Avanti's production facility and plumbed a new GM engine into the steel chassis and glassfibre body.

Blake even brought out a convertible in 1984, a full 20 years after the coupe first entered production. Financial difficulties however, ensued and the Avanti was then saved by Michael Kelly, a Texas magnate, who had the $750,000 handy to buy the remains of the company.

Kelly modernized the Avanti with a new dash, better interior and few slight styling tidy-ups. He also moved production to Ohio, and in 1987, brought out a limited-edition 25th Anniversary Avanti, in silver, naturally.

Kelly expanded on Blake's work bringing the range up to three models; with a four-door, a longer-wheelbase coupe and the rag-top. The 50 Anniversary cars were coupes, complete with supercharger pumping a good 250 bhp. Interiors were red or black leather and fully fitted out with TVs, telephone and sunroof. Alloy wheels were covered by fatter tyres and there were some body kit add-ons

The great thing about the Avanti is that despite being over 25 years old, it doesn't look its age. Styled sharply from day one, but consigned to the motoring scrapheap by a management lacking sufficient foresight, the Avanti has continued as a specialized, exclusive sports coupe – prices now come in just under £27,000 ($50,000). Since Studebaker sold on the moulds for the fibreglass body over 4,000 units have been made to personal order. If serious production had ever taken place, perhaps the car would have been developed out of existence by now and lost its classic appeal, so let's be thankful there are still a few newer cars left.

Avantis use GM running gear so spares shouldn't be any problem. Furthermore, fibreglass bodies resist rust and cope with light impacts. Interiors are to customer taste, so beware. However, once you've cornered an acceptable car, you'll have a very rare American commodity: a hand-built, crafted, personalized car, which isn't a stretch limo!

BELOW

*In peripatetic production for over 25 years, the Avanti has struggled to avoid becoming history for most of that time. The supercharged 25th Anniversary models produced 250 bhp and are the most desirable.*

## TECHNICAL SPECIFICATION

| | |
|---|---|
| **MODEL** Avanti Coupe/Convertible | **TOTAL PRODUCTION** c.4,000 |
| **ENGINE** 5001cc, V8 | **NUMBERS BUILT P/A** c.200 p/a in later years |
| **POWER** 250 bhp @ 4400 rpm (Anniversary Edition) | **ENGINE/CHASSIS NOS** N/A |
| **CHASSIS** Steel/fibreglass body | **SPARES AVAILABILITY** Unknown, running gear OK |
| **BRAKES** Ventilated discs/drums | **PRICE AT LAUNCH** $6,550 (1965); not officially exported to the UK |
| **TOP SPEED** 120 mph (192 kph) | **OPTIONS** Colour: Pearl silver<br>Trim: Red/black leather<br>OE Options: TV, CD, sunroof, supercharger, telephone, ground effect body styling |
| **ACCELERATION** 0–60 mph: N/A | |
| **PRODUCTION SPAN** 1962–1987 | |

## PAST CLASSICS

# STUDEBAKER 1950

Way back before Studebaker sold out the Avanti concern, the company had busied itself with taking on the big Detroit car makers in the immediate post war era. One of the results was the 1950 model complete with nose propeller, originally a gimmicky accessory, but latterly a prized possession for all '50 model owners.

Complete with 'next look' styling, which included a panoramic rear window and design detail reminiscent of a Dakota aeroplane, the propellor-nosed cars were sold with lazy straight six-cylinder 85 bhp engines, though later cars could be had with more powerful V8 motors. Even today 1950 prop-nosed cars can still be bought quite cheaply in the USA, and they will always make great American classics.

ABOVE
*The propellor nose may look outrageous but for those interested in the 1950 Studebaker, it's a must.*

U S A

# CHEVROLET CORVETTE ZRI

Corvettes have been around, firing the American sports imagination, for more than a quarter of a century, but none has created such a stir as the latest, fastest and most outlandish production 'Vette yet, the ZRI. Produced by General Motors, and drawing on the engineering talents of the recently acquired Lotus Cars concern, the ZRI is the car the Americans love to call the 'King of the Hill'. It's also the car most European critics have hailed as the 'car which out-Europeans the Europeans' in terms of performance refinement and character.

Lift the glassfibre bonnet and there's a traditionally sized 5.7-litre V8 in aluminium with four overhead camshafts, 32 valves and 380 bhp of usuable power lurking beneath the alloy cam covers. Sequential multi-port fuel injection and a healthy dose of Lotus Know-How get this 3454lb (1,568kg) monster moving all the way to 100 mph (160 kph) in 12 secs. The slightly larger-than-standard 'Vette body sits on a fully galvanized steel chassis, has flared arches to take the wider 315/35ZR rear tyres and four square, red rear-light lenses.

The ZRI started off being a bargain supercar if ever there was one, costing about $58,000 in the USA. Prices in the UK were first quoted at around £35,000 ($65,000), but swiftly escalated through £55,000 ($102,000) to around £90,000 ($167,000) by the time the cars came on stream. Only 3,000 will be made each year, and spreading them around the 4,600 US Corvette dealers will provide a headache, especially as overseas demand is substantial. This is one Corvette destined to be a classic example of American muscle in a refined package with universal appeal.

Rarity apart, the ZRI is developed with fast-paced touring in mind. The fully independent, electronically monitored and adjustable suspension has transverse front and rear leaf springs with a monoleaf made from a glass-epoxy compound. Suspension arms are of forged aluminium, and there's a limited slip differential to keep the power supplied to the wheels in contact with the road.

The ZRI drives easily; it isn't a car which needs brute force to hustle along at high speed. For such a heavy and powerful car to exhibit such un-American traits as nimbleness and sure-footed handling must indicate it is a car to be taken seriously in Europe. No boulevardeering in this 'Vette.

ABOVE

*The car they call 'King of the Hill' has power in abundance and a gimmicky key to lock out the top 180 bhp. Only 3,000 were built per year so demand is always strong. Prices rose fast initially , but are dropping back now.*

There's computer assistance with the manual six-speed gearbox, which automatically changes up if you accelerate too hard. 'Gas Guzzler' regulations ensure that in each gear the fuel consumption stays within Federal limits, though there is a rear wire, which if cut, it's claimed, deletes this helpful option.

If any cars come to the UK they'll be among the most prized American imports in decades. Character it may lack – you don't buy credibility with 'power keys' which lock out 180 bhp, leaving only 200 bhp for the days when teenage offspring borrow the car – but you do buy respect with the ultimate performance figures and handling.

## TECHNICAL SPECIFICATION

| | |
|---|---|
| **MODEL** Chevrolet Corvette ZRI | **TOTAL PRODUCTION** N/A |
| **ENGINE** 5727cc, LT5 32 valve, V8, twin port fuel injection | **NUMBERS BUILT P/A** 3,000 |
| **POWER** 380 bhp @ 6200 rpm/370 lb/ft @ 4200 rpm | **ENGINE/CHASSIS NOS** N/A |
| **CHASSIS** Steel frame/body: glass reinforced plastic | **SPARES AVAILABILITY** All parts available. |
| **BRAKES** Discs all round | **PRICE AT LAUNCH** $59,035 (£32,000); not officially exported to the UK<br>**OPTIONS** Colour: Red<br>Trim: Red leather<br>OE Options: Air bag, comprehensive computer, LCD graphics, cruise control, leather power seats, electric windows, FX3 selective ride control |
| **TOP SPEED** 180 mph (288 kph) | |
| **ACCELERATION** 0–60 mph: 4.2 secs | |
| **PRODUCTION SPAN** 1990– | |

PAST CLASSICS

## CHEVROLET CORVETTE STINGRAY

If you want raw power then look no further than the Corvette Stingrays of the 1970s. While they may share a visual similarity, the power produced by the big-block Chevy engines is to be believed only in the flesh. The hammering racket produced by the V8s on full song will stir the blood of the least interested of onlookers, and the cars were successful in winning races and sales for the more mundane, everyday V8-engined cars turned out by the thousand. The most exciting ran 7.4-litre engines capable of 460 bhp when stirred, and were available up to 1972. After the oil crisis, power was scaled down to under 200 bhp.

ABOVE

*The 1970 Corvette Stingray is typical of the hyped-up, muscle-car, power period. The hammering V8 engine managed 450 bhp plus, when revved hard enough.*

# DE LOREAN DMC 12

If you've seen *Back to the Future* you'll recognize the De Lorean. It's only a shame that the film makers didn't get around to carting Marty McFly back through time, because then the car's sales might have taken off as fast as the film's car moved through the night skies.

In reality, the stainless steel and glassfibre-bodied car had one of the most painful births in motoring history, dogged by under-funding and financial insecurity from day one.

By the time John Z De Lorean got the Dunmurry, Northern Ireland, production line rolling, having squeezed £67 million ($124 million) from the British government, the car was doomed. Build qualities were so suspect that the DMC 12 reportedly faced a total overhaul on reaching the USA, costing $2,000. Over 4,000 cars, 80 per cent of total production, ended up on that side of the Atlantic.

The car though was a stunner; gull-wing doors, unpainted, reflecting stainless steel body, and the Lotus Esprit-looking two-box shape, penned by Giugiaro, had presence. The slatted rear window created an air of mystery, though the car's overall looks suffered from the heavy Federal bumpers and rearward vision was minimal.

Power was supplied by the Peugeot/Renault V6, bored to 2850cc and slung behind the rear axle. Driving the rear wheels through a five-speed manual Renault 30 box (automatic was an option), the DMC 12 produced 140 bhp at 5300 rpm. Despite detractors prophesying awful on-the-limit handling with all the weight on the tail overhang, by cleverly fitting wider and larger rear tyres – 235-section as opposed to 195s at the front – the design hitch was miraculously cured.

There was a steel backbone chassis, which was attached to a fibreglass floorpan and superstructure. Stainless steel outer panels were then stuck on, and gull-wing doors fitted. The interiors were well laid out, with comfortable seats giving an excellent driving position.

With Detroit non too concerned if John De Lorean went to the wall, public interest, compounded by a mini recession in the USA, waned. Supply lines became fouled up and the banks closed in like hungry sharks on the trail of a feast. Refinancing deal after refinancing deal was wrung out of investors caught up in the De Lorean magic, but it was ultimately to no avail. A car which should have sold so much on its looks just wasn't given the chance it deserved.

Press criticisms of overweight, weakly powered cars carting a high price-tag did not bode well for a bright future, though today the unique combination of body style and finish, added to relative rarity, make the car a good long-term bet as an investment, and a sure-fire curiosity.

Bear in mind though, that for a similar price, a Lotus Esprit Turbo could be yours, and fewer of them have been built. They're still in production, and parts are readily available.

Still interested? Your next move should be to visit the USA, where cars are reportedly for sale at around £15,000 ($28,000) in a tidy state. You can be sure some wise investors have salted

BELOW
*This is one of only eight right-hand drive cars. The vast majority of the cars went to the USA. Good cars sell from £18,000 in the UK, $15,000 in the US.*

ABOVE
*History records that it failed miserably, a failure made all the more memorable by the massive investment by the British Government at a time when the home industry was desperate for cash. Gullwing doors brightened up an uninspiring bodyshell.*

brand new cars away, though at present, and for the short term, they will do well to cosset these unused cars, since they aren't quite ready to hatch yet.

Take care buying cars which appear cheap. Many were sold off practically for whatever the dealers were able to realize, and will have had equally uncaring owners. Low-value cars are best avoided, and cars which have had many owners should be especially shunned.

Rogue De Lorean cars aren't exceptions so it seems. Your best bet is non-US specified cars which won't have emission control suffocation as a no-cost option. Instead of the US car's 112 mph (180 kph) top speed, and non-existent 0–60 mph times, the European cars can top 130 mph (209 kph) and 0–60 mph in 9.0 secs, not blistering but sufficient.

## TECHNICAL SPECIFICATION

| | |
|---|---|
| **MODEL** DeLorean DMC 12 | **PRODUCTION SPAN** 1981–1982 |
| **ENGINE** 2850cc, V6, alloy, fuel injection | **TOTAL PRODUCTION** c.5,000 |
| **POWER** 140 bhp @ 5300 rpm | **NUMBERS BUILT P/A** c.2,500 |
| **CHASSIS** Backbone steel/stainless steel unpainted body fibreglass inner panels and floor | **ENGINE/CHASSIS NOS** N/A<br>**SPARES AVAILABILITY** Limited |
| **BRAKES** Discs all round | **PRICE AT LAUNCH** Up to $32,000 (£18,000); not officially exported to the UK<br>**OPTIONS** Colour: Stainless steel brushed finish<br>Trim: Leather<br>OE Options: Air conditioning |
| **TOP SPEED** 130 mph (208 kph) (UK Guide) | |
| **ACCELERATION** 0–60 mph: 9.0 secs | |

U S A

# 1988 PONTIAC FIERO GT

Real honest-to-goodness American two-seaters used to be about as common as a coherent foreign policy, but since the US government failed to outlaw sportscars without roofs, there's been a welcome resurgence in the popularity of sporting cars.

The Fieros were out to race with the Corvettes, trying to prove that even the blandest of shopping-cart mechanicals can be made attractive with a bit of careful styling and wodge of plastic.

The Fiero range appeared in 1984, complete with mid-engine four-cylinder power and stock Pontiac parts-bin running gear. Priced at $9,000 (£4,800), the Fieros were affordable, but underpowered. Answering calls for more go, a V6 found its way under the plastic engine cover, and eventually, by the time the market was ready and willing, a five-speed gearbox passed the power to the road wheels of the GT version.

Sales were, no doubt, helped by the Fiero gaining Pace Car status in 1984, and also by a series of careful styling fresheners during model-year up-dates. However, sales dived after the second year as complaints flooded in about lacklustre performance, hard ride, indifferent build quality and noisy engines. The smidgin of style within a two-seater, sleek-nosed shape couldn't help sales enough, and an official recall after a series of engine fires caused marketing nightmares.

By 1988 the Fiero was beyond salvation, despite an all-new front suspension, and GM sealed its fate with the halting of production. But for those after a brief look at a potentially exciting package the Fiero can deliver.

Go for a Pace Car replica from 1984 (limited edition of just 6,000), complete with body kit and drainpipe exhaust or failing that, the GT will do. The restyling gave the car a nicer rear-side profile as well as a five-speed gearbox.

The Toyota MR2, Fiat X1/9 and Fiero GT all share the same theme, and are pretty successful at it. As long as you accept there'll be little stowage space and relaxed performance, they are all cars to satisfy, especially the later GTs with their more specific suspension. While the Fiat has lighter steering, unusually for the American car the GT still had no power assistance by 1988 and twirling the wheel could be a chore.

RIGHT AND BELOW LEFT

*Pontiac Fiero GT promised so much, but engine fires ruined its reputation and Pontiac decided not to persevere. This is one of the better 1980s cars from the US.*

## TECHNICAL SPECIFICATION

| | |
|---|---|
| **MODEL** Pontiac Fiero GT | **TOTAL PRODUCTION** c.58,000 |
| **ENGINE** 2837cc, V6, OHV, fuel injection | **NUMBERS BUILT P/A** c.17,000 |
| **POWER** 140 bhp @ 5200 rpm/170 lb/ft @ 3600 rpm | **ENGINE/CHASSIS NOS** N/A |
| **CHASSIS** Steel 'X' chassis | **SPARES AVAILABILITY** Average USA availability |
| **BRAKES** Discs/discs | **PRICE AT LAUNCH** $11,795 (£6,300); not officially exported to the UK |
| **TOP SPEED** 115 mph (184 kph) | **OPTIONS** Colours: Red, white, black, blue<br>Trim: Cloth/leather<br>OE Options: Air conditioning |
| **ACCELERATION** 0–60 mph: 8.0 secs | |
| **PRODUCTION SPAN** 1985–1988 | |

PAST CLASSICS

## PONTIAC GTO

Pontiac were already exceptionally well versed in the art of classic car construction by the time they came up with the baby Fiero. The 1970s GTOs were US muscle cars supreme, now hot classics to collect. Powered by 6.5- and 7.5-litre V8s, at least 360 bhp was available. All steel GTOs separate chassis, a benefit for future classic restoration of the 50,000 or so which were produced between 1969 and 1971. Three body styles were offered with the coupe and hardtop far more numerous than the aggressive-looking convertible.

RIGHT
*1970s Pontiac GTOs were highly desirable muscle cars which are now hailed as classics. They are great value still, though cars with matching numbers, ie, original spec engines and chassis, are getting hard to find.*

# FORD THUNDERBIRD SUPER COUPE

The Thunderbird concept got a touch bogged down in the 1980s, overcome with emission control and doubtful body styling. Late in the decade there emerged a saviour, a throw-back to happier times of aggressive power and exhilarating driving. It was called the Super Coupe, but SC also stood for supercharged, that 1920s and 1930s method of blowing more power without resorting to increasing engine sizes.

With 1990s technology, the blown T-Birds now managed an impressive 215 bhp. Impressive, that is, when you realize the standard issue Thunderbirds were pushing out a mere 140 bhp on the same 232cu (3.8-litre) V6 blocks.

The 1990 cars bristled with technology too, from electrically selective damping and ride control, to digital displays and electronic ignition and injection systems. The SC took the best of these but hung on to analogue displays for the driver to peer at.

The shape was immediately loved. The US press was so taken with the car that *Motor Trend* awarded it the 1989 Car of the Year prize, quite an achievement considering the opposition. But it does look a shade too much like the now departed BMW 6 Series. It's the side-on view which gives the Teutonic appearance, the door and rake of the windscreen is rather too similar, or is it just a fluke of design?

Even so, the front and rear-end styling works hard to disguise the impact bumpers, and the only styling gaffe is the horribly over-heavy touches around the rearmost side window. To finish off the Thunderbird's stunning appearance a lovely set of styled 16-inch alloy wheels support wide, sticky tyres giving good grip.

Inside the new T-Bird there's a European flavour with a well set out dash and nicely trimmed seats; for 1991 the Thunderbird logo was incorporated into the design. Ford were deter-

BELOW

*The Supercoupe has a blown engine, which pushed out 215 bhp from its V6 3.8-litre engine. Succumbing to enthusiasts' demands, 1990 saw the reintroduction of a normally aspirated V8 engine.*

mined to make the SC a driver's car, and while it can't quite capture the flavour of the early two-seaters, the four-adult supporting body style of the 1990s does have a distinctive appeal for the press-on driver.

In the power department the 215 bhp will help the driver spirit the almost two-ton, two-door coupe to 60 mph in around 6.0 secs. Top speed, should he find an unpatrolled road, is in excess of 135 mph (217 kph), and acceleration all through the gears should be pretty exciting with the supercharger flexing its muscles. The advantage of supercharging over turbocharging is that the supercharger spins with the engine, so there's no lag as the turbo builds up and cuts in.

Buying a T-Bird SC in the USA should be simple enough, if you find the right dealer, though from what I hear, the cars are proving very popular. Definitely the most exciting Thunderbird for years, almost decades. The famed Ford V8 engine made a reappearance in the 1991 T-Birds, but not with a supercharger. Lazy power with V8s, crackling acceleration and spirit with the V6; the choice is yours.

BELOW RIGHT

*The Supercoupe has great lines, and is perhaps the nicest-looking Thunderbird for many years, reminiscent of the balcyon days of the 1950s and 1960s, when the T-bird looked its best.*

## TECHNICAL SPECIFICATION

| | | | |
|---|---|---|---|
| **MODEL** | Ford Thunderbird Super Coupe | **TOTAL PRODUCTION** | 23,710 |
| **ENGINE** | 3800cc, V6, supercharger, intercooler, elec. fuel inj. | **NUMBERS BUILT P/A** | 16,546 |
| **POWER** | 210 bhp @ 4000 rpm/315 lb/ft @ 2,600 rpm | **ENGINE/CHASSIS NOS** | N/A |
| **CHASSIS** | Unitary steel | **SPARES AVAILABILITY** | All parts available |
| **BRAKES** | Discs all round | **PRICE AT LAUNCH** | $19,823 (£10,715); not officially exported to the UK |
| **TOP SPEED** | 137 mph (219 kph) | **OPTIONS** | Colours: Standard Ford shades. Trim: Cloth. OE Options: LSD, ABS, power steering, automatic ride control, dual exhausts, CD player, power sunroof, special bodyside mouldings, vehicle maintenance monitor, supercharger boost gauge, electronic automatic temperature control |
| **ACCELERATION** | 0–60 mph: (NA) | | |
| **PRODUCTION SPAN** | 1989– | | |

PAST CLASSICS

## FORD THUNDERBIRD

Post-war American sporting car stylists took their lead from the European auto shows. When Ford designers saw the Jaguar XK 120, and heard arch-rivals Chevrolet were building a two-seater convertible, they set to work with the Thunderbird as the result. That was back in 1955.

Sales in the first three years amounted to close on 45,000 units. Demand was still strong for the V8-engined convertibles when the four-seater cars were introduced in 1958, T-Birds outselling Chevrolet Corvettes many times. Early 1950s T-Birds are still reasonable to buy, unless concours cars are on your shopping list. T-Birds are great classic American convertibles with their indespensible white wall tyres, though they lack the rust-resistant qualities of the glassfibre Corvettes.

# AUDI QUATTRO 20V

Developed from a 400-unit homolgation special, to serve as a test bed for their new four-wheel drive system, Audi's Quattro has been a phenomenal success. Indeed, the car is a phenomenon in its own right. From the time the first cars hit the streets in 1980 (they weren't sold in the UK until 1983), they proved an unstoppable success. The combination of the Audi's reliable and powerful 2.0-litre, turbo-charged five-cylinder engine and four-wheel drive in a sharp, modern, boxy shape was just what the market wanted.

After a string of rally victories (which stretched well towards the end of the decade on several continents), the two-door coupe gained universal acclaim. It brought Audi to world attention as the makers of well-engineered, thoroughly tested and remarkably fine motorcars.

LEFT
*Production of the Audi Quattro 20V was halted in 1991, after two years of producing one of the best value-for-money supercars on the road today.*

The Quattro body was also available as a two-wheel drive, normally-aspirated coupe, and sold for something like half the Quattro's price. This coupe too attracted volume sales, and the cars were so similar visually that many were able to pass themselves off as Quattro drivers.

Constant development has been the key to Audi's success with the Quattro. A short-wheel-base bespoke rally car emerged in the late 1980s, and these must be prized investments for the future. The longer, more sedate road cars, which were upgraded to 20 valve 220 bhp specification in 1989, complete with KKK K24 turbocharger and exhaust catalyst, are definitely entrants to the universal motor industry history books.

Despite rather bland styling overall, the performance package, coupled to the aura of rally competence and outright performance from the 200/220 bhp five-cylinder engines (depending on valve configuration), makes the Quattro a car to salt away for the years to come.

Ride and handling are very much above average, road manners and poise always creditable, and interior refinements, while occasionally bordering on the flippant, with LCD displays in the dash, served to keep the driver informed, and hold his attention on driving pleasure, not playing with gadgets.

The Quattro was the Grand Tourer of the 1980s, a car for the demanding but practical

LEFT
*The Quattro was the definitive early 1980s rally car, and the car which started the four-wheel drive performance revolution.*

driver who wanted the back-up of a volume manufacturer and the care of the specialist. Later cars with ABS and 20-valve 2200cc engines are the ones to look out for, but any Quattro which has been looked after and regularly serviced by a main dealer, and not driven hard in rally-driver style, will be a car worth preserving.

Prices of Quattros will always be steady, buoyed by demand and sound engineering. Replacement parts are widely available, and should be for the foreseeable future. They'll never be cheap, but for cars offering such superb driving, they're bargains compared to some supercars.

ABOVE
*The Quattro's squared-off looks have become well known, and with the new 20-valve engine, performance is as far ahead of the pack as ever.*

## TECHNICAL SPECIFICATION

| | |
|---|---|
| **MODEL** Audi Quattro 20V | **TOTAL PRODUCTION** 11,000 (all types) |
| **ENGINE** 2226cc, 5 cyls turbocharged | **NUMBERS BUILT P/A** 1,000 |
| **POWER** 220 bhp @ 5700 rpm/228 lb/ft @ 1950 rpm | **ENGINE/CHASSIS NOS** N/A |
| **CHASSIS** Unitary steel | **SPARES AVAILABILITY** All parts available |
| **BRAKES** Ventilated discs/discs | **PRICE AT LAUNCH** £32,995 ($61,000); NA ($28,750) |
| **TOP SPEED** 143mph (229 kph) | **OPTIONS** Colours: Standard Audi shades<br>Trim: Leather/cloth<br>OE Options: Metallic paint |
| **ACCELERATION** 0–60 mph: 5.9 secs | |
| **PRODUCTION SPAN** 1989–1991 | |

## PAST CLASSICS

# AUDI 100S COUPE

From 1969 to 1976 Audi produced another good-looking two-door, the 100S Coupe. Equipped with the 100-range 1.9-litre engine, the car's fastback body styling looked Aston Martin-like, though the top speed from the 115 bhp engine was a respectable, rather than heady, 112 mph (180 kph). Over 30,000 were made, all front-wheel drive. Good ones cost no more than £2,500 ($4,600), so they're practical, small sporty cars to buy. Watch for rust though.

RIGHT
*The Audi 100S Coupe is set to be a classic. While not overtly sporting, the shape has elegance and the car is refined to drive.*

# BMW 850i

Ignore, if you will, the price-tag, and feel the quality. The new 850i BMW is one of the more graceful, high-powered sports tourers to come from Europe in the past decade. Complete with six-speed manual gearbox, an electronics factory inside every centre console and wide door, and an odd mix of BMW M1 front-end looks, with currently *à-la-mode* rear end styling, the BMW has what it takes to be number one in Europe's super coupe market.

Silky smooth V12 apart, the two-door coupe makes pretentions of 2+2 accommodation, when it's really a two-seater grand tourer of the old school. Based on the thirteen-year successes of the BMW 6 series cars, the latest offerings attempt to win customers one notch further up the social scale.

Chassis development brought about the first coupe for decades which wasn't based on an existing saloon floorpan. The car encompasses several electronic innovations within the running gear – electronically controlled gearbox, multiplex wiring and active traction control, which prevents wheelspin when applying the rear brakes below 25 mph (40 kph) and gives a firmer grip.

The V12 engine is rarely troubled by hard work, as it dispenses more than enough power for wafting around town, and even autobahns won't make it sweat. Traditional to the last in running-gear layout, the 850i relies on proven manufacturing techniques and legendary BMW quality to see it through.

These cars feature such state-of-the-art electrial gimmickry as automatic window sealing when the doors close, electrical seats with memories, as well as seat belts built into the frames in a similar manner to Mercedes Benz's 500SL.

The high-brow executive market is in BMW's sights, and the car will sell on the blue and white badge as much as on its own charisma. For all the technical excellence, purring performance and genuine craftsmanship within, the V12 engine under the long sloping bonnet, capped with the famous kidney grilles, is what will get the sales going. The 850i is all alloy, and traditionally designed, with single overhead camshafts per bank, and two valves per cylinder, turning the car into a relaxed cruiser more than a street rod.

Finding a good car won't be a problem, since most will be specified as automatics, and will lead sheltered lives. Costing £25,000 ($46,000) more than the equivalent British offering, the Jaguar XJ-S V12 coupe, the car won't be a familiar sight for many, but it will be a rewarding car to own and drive. A car to collect for posterity, and an important one, as it's a flag-bearer of 1990s technology, and a good car to drive now.

BELOW
*The BMW 850i incorporates a silky smooth V12 engine but accommodation is strictly 2+2. UK order books are full until 1992, despite a price tag of almost £60,000 ($111,000).*

INSET
*Inside the 850i driver comfort is a priority and all electronic controls are to hand. Most cars will be automatics.*

| TECHNICAL SPECIFICATION | |
|---|---|
| **MODEL** BMW 850i Coupe | **TOTAL PRODUCTION** N/A |
| **ENGINE** 4998cc, V12, SOHC per bank | **NUMBERS BUILT P/A** c.10,000 |
| **POWER** 300 bhp @ 5200 rpm/332 lb/ft @ 4100 rpm | **ENGINE/CHASSIS NOS** N/A |
| **CHASSIS** Unitary steel | **SPARES AVAILABILITY** All parts available |
| **BRAKES** Ventilated discs all round, ABS | **PRICE AT LAUNCH** c.£60,000 ($111,000); $75,100 (manual) $75,450 (automatic) |
| **TOP SPEED** 155 mph (248 kph) (governed) | **OPTIONS** Colours: BMW's wide choice. Trim: Leather OE Options: Automatic stability control (automatics), M-technic sports suspension, Servotronic power steering (cost option) |
| **ACCELERATION** 0–60 mph: 6.8 secs | |
| **PRODUCTION SPAN** 1990– | |

PAST CLASSICS

## BMW 6 SERIES

Quite obviously, the BMW 6 Series cars have been well and truly eclipsed by the startling new 8 Series. But there's much on offer for BMW coupe fans in the old 6 Series. Go for Alpina conversions, or the Highline 1980s coupes. Fitted with full leather interiors, tuned engines and suspension packages, they're rare, swift and expensive.

Largely hand-built by Karmann in Osnabrück, the 6 Series cars were based on 5 Series floorpan, ran silky six-cylinder engines, and managed 0–60 mph in 7.3 secs without even trying. Top speed was never less than 140 mph (225 kph), and the hotted up Motorsport versions, the M635CSis, could muster 268 bhp and 0–60 mph in 6.3 secs. Available in a limited range of colours – racey red, silver, black and white (though there were others), the understated black or silver are the cars to buy. Prices for 6 Series BMWs are not too high but bag one now before the classic car world chases prices through their electrically operated sunroofs.

RIGHT
*Buy this car now! M635CSis won't be cheap for long, but are great value now that they've just gone out of production.*

# BMW M3 SPORTS EVOLUTION

Now don't get confused. This is the 1990, £34,000 ($63,000) car, not the Roberto Ravaglia Special Edition M3, nor even the M3 Evolution example from 1988. This is the furthest BMW had got with their 3 Series development at the time of writing and most probably as far as they'll get, remembering that a new shape '3' is on the way.

On the German race tracks the battle between the Mercedes 190s and BMW3 Series race-bred cars drove the search for more speed and cornering abilities into the realms of the ridiculous. As homologation demanded 500 road-going examples be built, there's a good number of lucky, wealthy enthusiasts out there with these on their driveways. In deference to the UK's constant rave over fast '3's, ten per cent of the limited-production cars will be sold in Britain.

The standard 3 Series combined elegance with engineering excellence. Increased power was always demanded, and in 1988 the first M3Es arrived with 10 per cent more at 220 bhp, and a top speed of 152 mph (243 kph). The sprint to 60 mph took 6.7 secs. It was special, available in red or black, and came complete with wide, racing, flared wheelarches and alloy-spoked wheels. Even better news for the BMW enthusiast – 180 super-special convertibles were produced, based on the M3E.

The next arrival was the Roberto Ravaglia special edition, in August 1989. He was the champion M3 driver who won the 1987 European Touring Car Championships for BMW. The UK received only 25 examples, each costing £26,850 ($50,000) and having bigger front and rear deck spoilers. More crucially, lightweight panels helped get this M3 to 60 mph in 7.1 secs, and it was fitted with a three way catalyst.

ABOVE
*The 238 bhp engine will send the car scurrying to 60 mph in 6.2 secs.*

Roberto's special editions were powered by the 4-cylinder 16-valve engine, with 170lb/ft of torque from the 215 bhp-tuned engine. Top speed was around 143 mph (230 kph), and there was a signed badge on the dash.

So now we come to the latest wallet basher, the BMW M3 Sports Evolution. Fifty were allocated to the U.K. With engines of 2467cc and 238 bhp and 177lb/ft at 4750 rpm, over 147 mph (236 kph) is possible, with 60 mph coming up in 6.2 secs. The extra power boosts the car in the upper reaches of the torque curve, while exceptional handling keeps the M3 Sports Evolution on the right line at almost any speed.

Adjustable spoilers for fine tuning of aerodynamic aids to keep the car glued to the road at high speed, varying the downforce when you move from motorway cruising to swinging through country bends.

Inside, amongst the leather, there are suede steering-wheel covers and gear knobs, and racing-style front seats complete with four-point harness slots. All M3s, even the cars built as souped-up saloons from 1986 on, are worth buying as classics. They are highly competent road cars, offering great performance, style and visual and driving appeal.

Later special edition cars are even more worthwhile, being exported in tiny numbers, reflecting the small overall build numbers. Coping with left-hand drive shouldn't be a problem, but expect insurance company nightmares.

## TECHNICAL SPECIFICATION

| | |
|---|---|
| **MODEL** BMW M3 Sports Evolution | **TOTAL PRODUCTION** 500 |
| **ENGINE** 2467cc, 16 valves, DOHC | **NUMBERS BUILT P/A** 500 |
| **POWER** 238 bhp @ 7000 rpm/177 lb/ft @ 4750 rpm | **ENGINE/CHASSIS NOS** AC79000 (Dec '89); AC79559 (Feb '90) |
| **CHASSIS** Unitary steel | **SPARES AVAILABILITY** See BMW for details |
| **BRAKES** Ventilated discs all round, ABS | **PRICE AT LAUNCH** c.£34,500 ($63,000); not officially exported to North America |
| **TOP SPEED** 147 mph (235 kph) | **OPTIONS** Colours: Red, black<br>Trim: Leather-trimmed interiors<br>OE Options: |
| **ACCELERATION** 0–60 mph: 6.2 secs | |
| **PRODUCTION SPAN** 1990 | |

## PAST CLASSICS

## BMW 2002 TURBO

Way ahead of its time in 1973, the 2002 Turbo was the best developed of the early small BMWs. Powered by a 1900cc four-cylinder engine, the 170 bhp catapulted the car to 130 mph (209 kph). Although almost 1,700 examples were created out of the stock 2002Tii shell in just ten months, the oil crisis made fast gas guzzlers unpopular. On the race tracks, turbo BMWs had been doing well for years, and in 1973, as the first European turbo road cars, they were curiosity items. Only 50 cars came to the UK, all left-hand drive, and hours of turbo lag to catch the unwary. Now priced well below £10,000 ($18,500) in a tidy state, they'll be sure to accelerate in value very soon.

RIGHT
*The BMW 2002 Turbo was the company's first Turbo road car and Europe's first small hot shot with a blown engine. The fuel crisis killed off the Turbo in its prime.*

# BMW Z1

BMW revived the sportscar with the Z1. It was the first real attempt by a major manufacturer to make a bold styling statement, with the emphasis on fast-fun motoring.

But underneath the high-tech plastic panels, which are easily unfastened and replaced, lies a conventional BMW 325i motor and gearbox. The independent suspension has a unique, cleverly designed rear set-up, which endows the diminutive sportster with a high degree of stability and safety.

Even with 170 bhp under the driver's right foot speed records won't be tumbling, but driver enjoyment levels will be high. The Z1 is eye-catching and good-looking. It has a cute style, reminiscent of the 'Frog-eye' Sprite, and the disappearing, electrically operated side doors, which sink into the sills: a gimmicky but useful piece of engineering ingenuity.

The Z1 needs no market definition. Priced in the UK at over £36,000 ($66,700) in late 1990, it's a rich man's plaything. Immediately identifiable, sizzlingly sporty, yet it runs on widely available and thoroughly proven mechanics.

A two-seater sports with simple-to-operate hood means you'll never have your spirits dampened, even if the heavens do decide to open. Wafting along with the BMW six-cylinder singing, doors fully wound down into their envelope, the Z1 could feel like a go-kart, and a very refined one at that.

The BMW Z1 is the definitive sportscar for the 1990s. It's sleek, striking and performs more than adequately for modern traffic conditions. Only six are built each day, and the queue for cars stretches into years. While running gear component parts will always be readily available, plastic body panels, and the composite fibre floor sandwich might be more difficult to source in later years, unless the Z1 starts a trend and further cars follow its lead.

RIGHT
*Bolt on plastic panels are an unique feature of the BMW Z1; a mix and match colour scheme is available for those who think they know better than BMW's decorators.*

## TECHNICAL SPECIFICATION

| | |
|---|---|
| **MODEL** BMW Z1 | **PRODUCTION SPAN** 1988– |
| **ENGINE** 2494cc, 6 cyls, fuel injection | **TOTAL PRODUCTION** c.6,000 |
| **POWER** 170 bhp @ 5800 rpm/164 lb/ft @ 4300 rpm | **NUMBERS BUILT P/A** c.2,000 |
| **CHASSIS** Steel monocoque chassis, fibre floorpan, plastic body panels | **ENGINE/CHASSIS NOS** N/A |
| | **SPARES AVAILABILITY** See BMW for details |
| **BRAKES** Ventilated discs/discs, ABS | **PRICE AT LAUNCH** £26,500 ($49,000); not officially exported to North America |
| **TOP SPEED** 136 mph (218 kph) | **OPTIONS** Colours: Red, yellow, metallics, black Trim: Leather. OE Options: Electronic mirrors and windows, alloy wheels, ABS, cen. lock., trip comp., air cond. (at extra cost) |
| **ACCELERATION** 0–60 mph: 7.9 secs | |

ABOVE
*BMW's baby sportscar made waves when launched and the high price did not deter queues of eager customers. It is a great car for the future, assuming you can find one.*

Definitely a classic, the ZI has the cheeky character people love in a sportscar, though to be fair, it is very expensive. The dramatically styled interior features well-padded bucket seats, with body-coloured plastic backs, and a roll-over hoop above the windscreen frame. Rear-view mirrors are sited a little way up the screen pillars, and the hood stows away beneath a plastic hinged lid, leaving a completely flat rear deck.

Demand for the car was so strong that black market prices shot up as high as £50,000 ($93,000). People foolish enough to spend that kind of money will have to wait many years to see returns on their investments, but will enjoy whiling away the time. As production is so restricted, cars will be in demand for a long time, so the best way to acquire one of these mouth-watering cars is to join the waiting-list.

PAST CLASSICS

## BMW M1

Designed for the track, the BMW M1 appeared in 1977 complete with mid-mounted 3.5-litre 24-valve, six-cylinder turbocharged engine.

Its fibreglass body was mounted on a steel frame. The M1 was developed to win on the race tracks and sell BMWs to the man in the street, but despite the enormous cost at the time the cars were never as successful as the Bavarian company hoped. Top speed was posted at around 160 mph (258 kph), with the sprint to 60 mph achieved in a mite over 6 secs. The M1 engine produced 277 bhp @ 6500 rpm, but plans were made to boost that to almost 800 bhp for racing purposes.

BMW had to build 400 examples for homolgation for racing in the late 1970s and 456 were eventually built for sale during 1979 and 1980. With so few cars built, rarity increases their value. Furthermore, being such attractive, easily identifiable sports coupes, M1s will never be cheap. Today cars will fetch £100,000 ($185,000) with ease, and despite the rather garish colour schemes with which some were originally endowed, any M1 is sure to be a classic investment so long as it's in good mechanical and structural order.

RIGHT
*The racing pedigree of the M1 shows through on the road. Softer springing to improve the ride comfort does not deter from the car's natural poise and stability.*

# PORSCHE 959

Surely this is the definitive 'Oh my Gawd' supercar: the car which does everything. It's got 450 bhp under the rear hatch, leech-like four wheel drive, and Porsche-Motorsport derived wheels and tyres to put the power down. It's got a body and chassis made up of a central steel tub, with aluminium doors and opening panels for lightness and a polyurethane nose with the rest of the body being a Kevlar composite.

Only 200 or so 959s ever rolled out from the specially built Porsche production line. All went to pre-arranged, delighted owners. To ever see one on the road would be a major achievement, and to drive one would put you in a very exclusive category indeed.

Not that driving the 959 demands exceptional skills; four-wheel drive banishes rear-wheel steer experienced in standard 911s, and the widened track and fatter tyres keep the body-roll to an absolute minimum. Within the gearbox there are six ratios, one designed to take the car over rough terrain, not that many owners will be crossing too many ploughed fields.

The 959 was designed with Group B competition in mind, to run against the likes of Ferrari's GTO. Sadly, the stand-off never took place, as the grouping was banned.

Production only began once there was a large enough queue of people, brandishing fat cheque books. Once this happened the Porsche people began to realize that the 959 would earn them good money, as well as enhance their image the world over.

The car features advanced electronics, which control not only the torque split between front and rear axles (depending on load and conditions under tyre), but also the ride height. At around 100 mph (160 kph) the body settles lower, enhancing already excellent aerodynamics, and with the appropriate spoilers, the 959 exhibits no lift at all at high speed. Vents scattered all over the body allow for engine and mechanical cooling.

Numbers alone guarantee the 959 classic status, but despite being warned off the field of play even before it had to show its racing poten-

BELOW
*Four-wheel drive gives secure grip on bends, ride height is speed sensitive, and the body doesn't suffer from lift at any speed, so Porsche claims.*

## TECHNICAL SPECIFICATION

| | |
|---|---|
| **MODEL** Porsche 959 | **TOTAL PRODUCTION** 200 |
| **ENGINE** 2849cc, DOHC flata 6, 24 valve, twin turbos | **NUMBERS BUILT P/A** c.50 |
| **POWER** 450 bhp @ 6500 rpm/369 lb/ft @ 5500 rpm | **ENGINE/CHASSIS NOS** 1987 – 95JS900213; 1988 – 95JS900284 |
| **CHASSIS** Steel/body aluminium/Kevlar | **SPARES AVAILABILITY** Ask Porsche |
| **BRAKES** Discs all round | **PRICE AT LAUNCH** c.£100,000 (c.$185,000); not officially exported to North America |
| **TOP SPEED** 190 mph (304 kph) | **OPTIONS** Colours: Red, silver<br>Trim: Leather interiors<br>OE Options: Engine tuning packages for racing or higher performance road use |
| **ACCELERATION** 0–60 mph: 3.7 secs | |
| **PRODUCTION SPAN** 1985–1988 | |

tial, the 959 is revered as one of the finest cars ever built. As one would expect from the Porsche engineers the workmanship is excellent.

No self-respecting multi-millionaire should be without one, but it's unlikely that many will spend much time actually eating miles at over 190 mph (304 kph). Most spend their days tucked up, rapidly appreciating in value. Who could bear destroying such a vehicle in an accident?

Twenty motorsport-prepared cars are supposed to have been readied for sale to aspiring wealthy racers, but with the demise of Group B, they were made redundant. These cars, with their highly tuned engines and racing aspirations, are the best 959s to acquire for any serious collection. Any one of the three different spec. and trim packages offered by Porsche is worth buying as an outright investment.

PAST CLASSICS

## PORSCHE 911 4×4 (1984)

Both the current Carrera 4 and the 959 share four-wheel drive technology although this concept had been under investigation by Porsche for many years, prior to its adoption for these two classics.

The 911s, sponsored by the Rothman's Rally Team for the 1984 Paris-Dakar Rally, used the standard Carrera 2 3.2-litre motor and a four-wheel drive system, complete with connecting torque tube linking front and rear axles. The team of three cars successfully completed the gruelling trans-continental desert dash, finishing an impressive 1st, 6th and 26th. The cars utilised several components borrowed from the Turbo, especially in the suspension department, and ran a raised body height of over 10 inches above that of the standard road cars.

Complete with two fuel tanks, one in the cockpit, team leader Jackie Ickx made the 1984 a memorable Porsche victory. Three years later in 1987 a Porsche 959 won the Paris-Dakar Rally, again in Rothman's colours.

RIGHT
*The Porsche 911 4×4's crowning glory came in 1984, when it won the infamous Paris–Dakar Rally with Jackie Ickx in the cockpit.*

# PORSCHE 928GT

The Porsche 928GT is yet another fine development of a superb theme. The 928 emerged way back in 1978, and evolution has been slowly turning this fast, refined road car into a hotter property ever since. Now boasting 330 bhp in this 5.0-litre V8 1989 guise, the GT combines the capability of an out-and-out tourer with the docility of a town car. Provided with an almost practical hatch, the 2+2 coupe's squat body knows how to eat up the miles.

Always the Porsche flagship, the 928GT grew out of the popularity of the Series 4 'Sport Equipment' cars of 1988. Uprating the engine and handling package even further, turned the GT into the most perfect sporting coupe ever produced by Stuttgart. Casting aside the competent and oft-chosen automatic box, the GT runs through a manual five-speed and limited slip differential, and sits on wide 225/50ZR16 and 245/45ZR16 Bridgestone tyres supported by stiffer-than-normal gas-filled shock absorbers.

Porsche's unique tyre pressure monitor is standard and there's ABS and an ultrasonic burglar alarm included in the GT's list of desirable features. Apart from these changes, the 928S4 and GT are very much the same, but you'll spot the GT's twin stainless steel exhausts, central locking, powered 'Sport' seats and cruise control.

The 928 vied closely with the Jaguar XJ-S as the Grand Tourer of the 1970s, and both still soldier on today. There's no let-up in production, and refinement is an on-going process for both. While there's no convertible 928, sales have continued solidly as the car is a high-performance package with status.

As the flagship of a company with such a flawless prestige image, the 928 would sell at twice the price, so the bargain-hunting classic investor won't find many cars. Practical to keep if you've a friendly Porsche dealer, the 928 is actually quite user-friendly, though many have been well run in as high-speed, motorway express cars.

The 928GT is the crowning glory of the current range, the Series 4 cars having more dramatic looks and better interiors than earlier cars. While early 1980s examples will fetch anything from £10,000 ($18,500) upwards, the GT will always be that almost unobtainable, and highly prized 928 everyone will want for its tremendous performance and luxury fittings.

Any 928 is a special car to savour, and a good classic to own in terms of engineering competence and radically different styling. Nothing looks or goes quite like the 928.

## TECHNICAL SPECIFICATION

| | |
|---|---|
| **MODEL** Porsche 928GT | **TOTAL PRODUCTION** 1,366 to end 1990 |
| **ENGINE** 4957cc, V8, DOHC per bank, 32 valves, 3-way catalyst | **NUMBERS BUILT P/A** 1989 – 369; 1990 – 997 |
| **POWER** 330 bhp @ 6200 rpm/311 lb/ft @ 4100 rpm | **ENGINE/CHASSIS NOS** 92LS840001 – 92LS860506 |
| **CHASSIS** Unitary steel | **SPARES AVAILABILITY** All parts available |
| **BRAKES** Discs all round | **PRICE AT LAUNCH** £55,440 (c.$103,000); $74,545 (1990) |
| **TOP SPEED** 170 mph (272 kph) | **OPTIONS** Colours: Standard Porsche shades<br>Trim: Leather<br>OE Options: Sports seats, ABS, tyre pressure monitor, ultrasonic alarm |
| **ACCELERATION** 0–60 mph: 5.8 secs | |
| **PRODUCTION SPAN** 1989– | |

ABOVE
*With over 330 bhp on tap, the V8 928GT is a 1990s muscle car, though docile in town driving. Expensive both new and second hand, the GT will be a popular version for many years to come.*

PAST CLASSICS

## PORSCHE 356B

The 911's grand-daddy, the 356B, managed 110 mph (177 kph) and 0–60 mph in under 10 secs'. Apart from being available in a multitude of body styles, from hardtop to cabriolet, the 356B was sold in great numbers, over 30,000, with many crossing the Atlantic to North America. The earlier 356A only had 700 or so siblings, but the most valuable 356 currently is the 356B/C Carerra fixed head, which sells for around £60,000 ($111,000) in concours form today. Regular 356Bs sell for as little as £20,000 ($37,000).

RIGHT
*The familiar 'Porsche nose' of the previous 356A was changed to a much prettier American bow with high-mounted bumpers and large overriders.*

# PORSCHE CARRERA 2/4

Yes, this is the one with the pop-up rear spoiler; the outrageous spoiler on the engine cover of past models has been banished. Re-creating the world's best-loved rear-engined sportscars for the 1990s was a task not lightly undertaken by Porsche, who wisely decided to offer two cars; one for the fraternity desperate to show off a four-wheel drive capability in the car park; and a simpler, nimbler two-wheel drive classical version, for those more intent on getting from A to B without fuss or drama, but with equal style and motoring satisfaction.

The Carrera Series 2 models are a timely redevelopment of the traditional, classic 911 theme. However, with extra-fine engineering input, the old 911 vices have been designed out. The four-wheel drive splits the power so that you can pull yourself through corners, and ABS removes the tendency for the light front end to lock its wheels under heavy braking in the wet.

The cars are identical save for the transmission systems. The new Carrera comes in three traditional flavours, coupe, targa and cabrio. The cabrio has electric help for those too feeble, or too lazy, to erect the fabric hood themselves.

Inside, the little-changed interior does now boast electric seats, but the dash is the same, and the knobs and switches lie about the place much as they did before.

The driving impression still reeks of speed and performance, the four-wheel drive system giving much more in adverse conditions, although requiring slightly more effort. In either case the car is a perfect example of delightful motoring for the well-heeled, who want to listen to the symphonic sound of a busy, flat six-cylinder engine. For the really laid back, there's a radical new 'Tiptronic' automatic gearbox, which, as you'd expect, is electronically controlled through five differently mapped programmes. Apart from the automatic function, you can select to drive it as a semi-automatic/semi-manual, by pushing the gear lever into a separate gate.

The new 911s will appeal to the sporting driver more and more through the 1990s. Not that all 1980s 911s should be written off as pretentious driver's cars; its just that too many of them fell into the wrong hands. Carrera 2 cars bask in the glow of year-long waiting lists, and this demand will hold up for many moons.

BELOW
*Revamping the old 911 for the 1990s revealed another sleek, pleasing shape, and with the addition of four-wheel drive in the '4' model, grip is more tenacious than ever. The Carrera 2/4 carries through the timeless 911 character, though now sporting better aerodynamics and more goodies, including the new tiptronic automatic selectable gearbox.*

Porsche cleverly prolonged the life of their most famous and numerous product, carefully styling it to meet the trends and demands of the 1990s, a two-way catalyst is standard for instance. They sensibly looked back to the purer days of the 1960s for the styling revitalization, away from the ugly, wide wheelarch monsters with flicking picnic tray-spoilered tails, back to the simple svelte shape. Electrically operated aerodynamics, which are seen, but don't scream, complete the picture.

## TECHNICAL SPECIFICATION

| | |
|---|---|
| **MODEL** Porsche Carrera 2/4 | **TOTAL PRODUCTION** Carrera 2 12,453; Carrera 4 17,723 |
| **ENGINE** 3600cc, flat 6, twin plugs per cylinder | **NUMBERS BUILT P/A** 1989 9,507 (Car 4); 1990 12,543 (Car 2) |
| **POWER** 250 bhp @ 6100 rpm/229 lb/ft @ 4800 rpm | **ENGINE/CHASSIS NOS** 96LS400001 – 96LS472169 |
| **CHASSIS** Unitary steel | **SPARES AVAILABILITY** All parts available |
| **BRAKES** Ventilated discs all round, ABS | **PRICE AT LAUNCH** £41,505 (2), £47,699 (4); 2 $60,200 (1990), (4) $70,400 (1990) |
| **TOP SPEED** 174 mph (278 kph) | |
| **ACCELERATION** 0–60 mph: 5.7 secs | **OPTIONS** Colours: Standard Porsche shades<br>Trim: Leather<br>OE Options: Air conditioning |
| **PRODUCTION SPAN** 1989– | |

## PAST CLASSICS

## 911 TURBO (1975–77)

The 911 Turbo grew out of the factory's works racing team; a blown 911, complete with massively extrovert rear spoiler. The motivating power was the flat-six engine, bored to 2944cc and reaching 260 bhp. Developed without a five-speed box and any other body styles, 2,873 Turbos were built between 1975 and 1977 and could all better 150 mph (240 kph). Handling was traditonal 911 and liable to catch out the inexperienced driver, with even more care needed in the wet. These models have exceptional classic potential with prices already stratospheric. After 1977, the engine size became 3.3 litres and significally greater numbers were produced.

RIGHT
*The Porsche 911 Turbo used a 3-litre motor at first. The tea-tray spoilers were introduced later; along with flared arches at the back.*

# VW GOLF LIMITED

If you're ever tootling along and see a VW Golf with an unfamiliar blue piping inlay in the front grille appear in your rear view mirror, prepare to pull over. You'll soon be passed by one of the fastest Golfs ever made for the road. Forget the GTi and 16-valve GTi, the Golf Limited is the fastest car in its family. Only 70 were built, a programme which grew out of an excercise to see how a G60-supercharged engine would fit a road car, which also carried the Synchro four-wheel drive system.

The base car is a standard 139 bhp 16-valve GTi, a four-door model. Each car goes from the production line to VW Motorsport in Hanover, where in time, the Synchro four-wheel drive system gets grafted into the car. Add in the Passat GT's gearbox, uprated brakes, 15-inch wheels with 235 section tyres, a catalytic converter and a supercharged and intercooled Turbo G60 engine and you're on your way to the finished product.

Uprated suspension keeps the Golf cornering with the legendary grip. Leather bucket seats hug the driver's and passenger's bodies, as well as a fully specified electric package inside the car, windows and mirrors etc. Then take a deep breath and say £22,000 ($40,750).

VW Motorsport are only seriously busy for part of the year. Rather than lie around idly ticking over, they've come up with both the Limited Golf and a great project to turn 20 newly discovered, brand new Beetle bodies into Karmann Cabriolets, selling each one even before it was built. It's the same sales success story as the Limited.

The Limited appeared out of the Golf Turbo project, a concept car which initiated 400 firm orders. VW saw the market and designed the Limited to fit. Externally, unless you've a sharp eye to spot the wider tyres, the car looks like any other 16-valver. Slightly flared arches give

BELOW
*The Golf Limited is the proverbial wolf in sheep's clothing. Only the blue grille insert gives the game away, though ardent Golf spotters may notice the bigger wheels and tyres.*

the game away a little more to the seasoned observer, but once you see it on the move you'll know it is no ordinary Golf.

Only one car, a dark grey one, came to the UK, and it was a left-hand drive model. It will be surprising if more than a couple of others follow and all cars will be left-hand drive. In theory, Germany is comparatively bursting with the other 69, yet to prise one away from an eager owner will take many Deutchmarks. The next best thing is to buy a Rallye Golf, or the more up-spec SE version. Costing around £18,000 ($33,000) when new, they'll satiate most people's appetite for speed and traditional Golf handling.

Obviously the Limited and Rallye cars are far better classic investments than the standard Golf GTis, which are just too numerous. Without a doubt, in say ten years time, the light, well-balanced Mk I cars will be in demand, and cars in good condition that have been well-looked after will be worth money.

If you want an Mk I GTi buy now, while they're at the bottom of their depreciation slope; as little as £1,500 ($2,800) should get you behind the steering wheel. Later Mk 2 Golfs are less desirable, except in G60, or 16-valve form, though this last car is less of a mind-blower than one might expect.

## TECHNICAL SPECIFICATION

| | |
|---|---|
| **MODEL** Volkswagen Golf Limited | **TOTAL PRODUCTION** 70 |
| **ENGINE** 1800cc, supercharged, 16 valves, twin catalyst | **NUMBERS BUILT P/A** 70 |
| **POWER** 210 bhp/186 lb/ft @ 5000 rpm | **ENGINE/CHASSIS NOS** N/A |
| **CHASSIS** Unitary steel | **SPARES AVAILABILITY** Body and running gear can be sourced |
| **BRAKES** Ventilated discs all round, ABS | **PRICE AT LAUNCH** £22,570 (c.$40,750); not officially exported to North America |
| **TOP SPEED** 142 mph (227 kph) | **OPTIONS** Colour: All 70 cars gunmental grey<br>Trim: Dark leather. OE Options: Heated front seats, leather interior, on board computer, central locking |
| **ACCELERATION** 0–60 mph: N/A | |
| **PRODUCTION SPAN** 1989–1990 | |

## PAST CLASSICS

## KARMANN GHIA 1968 CONVERTIBLE

The Karmann Ghia brought the VW Beetle acceptability on the surf-covered beaches of southern California. Based around the Beetle's running gear, the Karmann Ghia convertible was a smartly sophisticated two-seater which loved to play. While handling and performance was way off Porsche's territory, in the looks department it was the tops. Karmann Ghias love to rust, and panels are expensive, but with few sound examples readily available and prices rising, a spot of restoration could be the way to get hold of a fuss-free classic at a reasonable price. Over 80,000 convertible Mk1s were sold in all, compared with more than 350,000 coupes.

RIGHT
*Although its looks always distinguished the Karmann Ghia from the other run-of-the-mill vehicles of the day, only recently has it achieved classic car status.*

# INDEX

**Note:** References in italics are to illustrations, but there may be textual references on the same page.

# PICTURE CREDITS

J Baker Collection pp: 6 top, 9, 11 bottom, 12, 13, 17, 25 bottom, 26, 27, 29, 30 top, 31, 32, 33 top, 34 top, 37, 39 top, 43, 45 top, 48, 49 top, 51 top, 54–56, 61–66, 68 top, 74–77, 81 bottom, 83 bottom, 85–89, 91 bottom, 93, 95, 100, 101, 106, 107 top, 111, 114 top, 117, 124, 125

Alfa Romeo pp: 68, 69 bottom, 70, 71 top, 72, 73 top
Alpine Renault pp: 59
Andrew Morland pp: 33 bottom, 45 bottom, 53, 70 bottom, 96, 103, 113
Aston Martin pp: 10, 11 top
Bentley Motors Ltd pp: 14
Bertone: 84
Brian Snelgrove pp: 67
BMW pp: 108–110, 112, 113
Chevrolet pp: 98, 99
Christies pp: 6 centre
Citroen UK pp: 60
Daimler Benz AG pp: 114–116
Ferrari pp: 7, 8, 78, 80, 81 top, 82
Ford Motor Co pp: 16–19, 104, 105
Haymarket Magazines pp: 73 bottom, 97
Honda UK pp: 46, 47 top
John Howie pp: 47
Jaguar Cars Ltd pp: 20–25 top
Lancia (Fiat SpA) pp: 90, 91 top, 92
Lotus pp: 28, 30
Maserati pp: 94, 95 top
Mazda Cars UK pp: 49 bottom, 50, 51 bottom
National Motor Museum pp: 15
Nissan UK pp: 52
Panther Car Co pp: 34 bottom, 35–37
Pininfarina pp: 79, 83 top
Pontiac pp: 102, 103
Porsche UK pp: 8, 118–123
Renault UK pp: 58
Rolls Royce Motors pp: 40, 41
The Rover Group pp: 38, 39, 42
Toyota UK pp: 57
TVR SEAC pp: 44
VW Audi GmbH pp: 107 bottom